1979

Genetic Engineering

Laurence E. Karp, M.D.

Genetic Engineering:
Threat or Promise?

Medical Illustrations by Jan Norbisrath

Nelson-Hall nh Chicago

Library of Congress Cataloging in Publication Data

Karp, Laurence E.
        Genetic engineering, threat or promise?

        Bibliography: p.
        Includes index.
        1. Genetic engineering — Social aspects. 2. Genetic
engineering.   I. Title.   [DNLM: 1. Genetic intervention.   QH442 K18g]
QH442.K37   301.42'3   76-3497
ISBN 0-88229-460-1

Manufactured in the United States of America

TO

Dr. Joseph Seitchik and Dr. Walter Herrmann

who taught me that the wise man is
usually neither the first nor the last
to exchange an old idea for a new one

For in much wisdom is much vexation;
And he that increaseth knowledge increaseth sorrow.

Ecclesiastes 1.18

Contents

Acknowledgments   ix

Preface   xi

INTRODUCTION

Principles of Genetics   3

Part I   GENETICS IN DIAGNOSIS AND THERAPY OF DISEASE

1   Genetic Counseling   35

2   Eugenics, Euthenics, and Euphenics   49

3   Genetic Screening   65

4   Genetic Therapy   85

5   The Prenatal Diagnosis of Genetic Disease   107

Part II   REPRODUCTIVE ENGINEERING

6   Artificial Insemination   131

7   Sex Determination   149

8   Ectogenesis   161

9   Parthenogenesis   185

10   Cloning   195

11   The Synthesis of Life From Nonliving Matter   207

Notes   215

Glossary   217

References   223

Index   231

Acknowledgments

I'm grateful to many people for helping me to complete this book.

I thank my colleagues in the Center for Inherited Diseases at the University of Washington School of Medicine for their forebearance as I constantly raided their minds, their libraries, and their photo files. These persons include Drs. Arno Motulsky, Roger Donahue, Judith Hall, James Detter, Holger Hoehn, George Martin, C. Ronald Scott, David Smith, and Philip Fialkow, and Mrs. Jean Bryant. Other faculty members at the University of Washington who helped me with references or photographs were Drs. Ron Lemire, Thomas Standaert, Mary Forster, and Sue Conrad.

Thanks are also due to W. Dianne Smith for the fine karyotyping and the many excellent original photographs she prepared for me, and to Dr. Holger Hoehn for furnishing the chromosomal preparations which were used for most of the karyotypes.

Sonia Watson, Dr. Arno Motulsky, and Dr. Roger Dworkin were kind enough to criticize the manuscript and offer useful suggestions with regard to constructive alterations.

I'm deeply indebted to Barbara Erikson for typing the manuscript. In addition, my wife Myra provided helpful and necessary general secretarial assistance.

Preface

Many people today regard the subject of Genetic Engineering as threatening at best and thoroughly frightening at worst. Our newspaper columnists, book writers, and movie producers leave us with the unmistakable message that failure to stop the mad genetical researchers will send us irreversibly down the road to dehumanization, atrocity, and slavery.

We're told that the governmentally-controlled reproductive practices of *Brave New World* are right around the corner: babies will be mass produced to specification in test tubes, cloned into existence from skin cells of already-living adults, or even bred from artificially activated unfertilized eggs, thereby rendering males obsolete. In addition, we read about the major increase in homosexuality that will follow upon the application of techniques to permit preconceptual sex determination of fetuses. We hear that on our Sunday afternoon strolls, we'll soon be able to say hello to cyborgs, human-animal chimeras, and specially-bred legless astronauts. Supposedly, genetic screening will lead to caste assignations based on the goodness or badness of our hereditary material. And talk is thrown out concerning the possibility of Nazilike attempts to "eliminate the unfit."

So much for the threat. We don't hear a lot about the other side of the coin: the promise is nowhere near as sensational. But it's likely to be much more significant. Through the application of genetic knowledge, we may achieve true breakthroughs in the prevention and alleviation of birth defects and other human genetic disorders, heart disease, and even cancer. Genetic know-how might provide the means to feed the millions of hungry people in our world.

Are things beginning to sound a little better now? Or are the issues just getting more confused? How can you tell what and whom to believe?

These questions bring me to the basic reasons for writing this book, which is neither a general textbook of genetics nor a treatise on contemporary moral and ethical issues. In a sense, it is a how-to book, a manual to help the layman understand the nature and implications of current and future developments in Genetic Engineering. Each chapter will cover a different area of Genetic Engineering; I present and discuss the controversial aspects of these areas. But it's not sufficient simply to be aware of the debatable issues. In addition, one must know and understand the basic facts and principles of Genetic Engineering concepts and techniques. Only then does it become possible to reach truly informed decisions regarding the propriety of the development and use of Genetic Engineering. For this reason, each chapter in the book contains a thorough explanation of the concepts behind the controversies.

At this point, it seems appropriate to consider the primary concept. What is Genetic Engineering, anyway?

Working from the American College Dictionary's definition of "engineering," we can extrapolate to: the art or science of making practical application of the knowledge of the pure science of genetics. Thus, we might consider Genetic Engineering to represent any attempt to modify the structure, transmission, expression, or effects of the genes, the ultimate directors of heredity.

A definition of Genetic Engineering printed in the *Journal of the American Medical Association* in 1972 reads in part as follows:

> The popular term, genetic engineering, might be considered as covering anything having to do with the manipulation of the gametes (reproductive cells: sperm and eggs) or the fetus, for whatever purpose, from conception other than by sexual union, to treatment of disease *in utero,* to the ultimate manufacture of a human being to exact specifications. It has nothing to do with the creation of life; it is concerned only with the method for transmitting life.[1]

In fact, this statement appears to define Reproductive Engineering, an area which I would consider to comprise only one of the two major fields of Genetic Engineering. Reproductive Engineering includes artificial insemination, pre- and post-conceptual sex determination of offspring, *in vitro* fertilization and ectogenesis (conception and fetal development outside a mother's body), parthenogenesis ("virgin birth"), and cloning (reproduction using genetic material from cells other than gametes). In addition, despite the disclaimer in the AMA definition, I believe that the synthesis of new life from nonliving material should also be included under Reproductive Engineering.

The second major area of Genetic Engineering embraces those endeavors that involve the application of genetic knowledge to the more traditional medical problems of diagnosis, prevention, and treatment of disease. Basic to all other fields in this sphere is genetic counseling, which is the process of making accurate diagnoses of genetic diseases, followed by explanation to the patients and their relatives. Procedures that may properly follow upon genetic counseling are eugenics (genetic manipulations designed to improve health), euthenics (environmental manipulations to the same end), screening of populations to detect or prevent genetic diseases, specific therapy of genetic diseases, and prenatal diagnosis of genetic diseases.

An often-used term is "The New Genetics." This is a catch phrase, usually loosely equated with Genetic Engineering. It has no more real meaning than "The Old Genetics," or "The New Psychiatry."

The controversies related to Genetic Engineering go a long way toward explaining the recent proliferation of specialists in bioethics. These individuals are drawn from the fields of biology, theology, philosophy, and law. They define and discuss moral and ethical problems associated with the practice of medicine, including the genetical issues of the day. Sometimes, their conclusions lead them to call for regulation or suppression of further work in different areas of genetics. I think this is unfortunate, because morality systems are totally subjective, arbitrary, and changeable according to the consensus in different locations and at different times. A professional bioethicist may be thoroughly conversant in history, philosophy, and religion, but he possesses no special facility for deciding for anyone but himself what is the proper moral or ethical attitude to adopt in a novel situation. Inevitably, he must put forth his own views and opinions, or those of his sponsors; these are by no means in accordance with or more reasonable than your ideas or mine. Legislation of morality is acceptable only when it's your own particular brand of morality, and it's not a large step from permitting others to decide which books are fit to be read or whether we may legally take a drink of liquor, to abdicating our right to decide which forms of knowledge we may lawfully pursue, acquire, and utilize. Forbidding people to use genetic knowledge and forcing them to use it seem equally wrong to me. Therefore, I believe education to be necessary, so that the layman may acquire the factual background he needs to reason logically and intelligently, and then to act within the bounds of what he considers right and proper. In this way, the option to use or refuse Genetic Engineering can remain with the individual.

As mentioned, by far the larger proportion of the Genetic Engineering debate has focused on the threat, rather than on the promise. Two of the leaders of the opposition to unrestricted genetic research are biologist Leon Kass and theologian Paul Ramsey. Their position is based upon convictions of a metaphysical nature. Kass' arguments arise primarily from two principles. The first is the belief that a conceptus is a human being and further, that it is unable to give informed consent. Therefore, Kass sees Reproductive Engineering as constituting "unethical experiments on the unborn" (actually, he means "unconceived"). His second thesis centers on his conviction that we are not "wise" enough to be able to predict the outcome of our genetic explorations. This being so, he concludes:

> In the absence of that "ultimate wisdom," we can be wise enough
> to know that we are not wise enough. When we lack sufficient wisdom
> to do, wisdom consists in not doing.[2]

Kass' papers are highly rhetorical: for example, he regards any attempts to "remake ourselves" as representing the "folly of arrogance." Sometimes, he borders on the hysterical, as when he considers the possibility of the production of 10,000 Mao Tze-Tungs by cloning, or expresses the fear that the "new technologies" may result in the replacement of humans by nonhuman forms of life.

Paul Ramsey reaches the same negative conclusion as Kass. He also holds firmly to the belief that any conceptus of any gestational age must be considered a human being. In addition, he sees attempts to alter man's basic genetic makeup or to create life by novel techniques as unacceptable efforts to defy divine mandates and usurp divine prerogatives.

This approach, then, is based upon arbitrarily defined, rigid moral laws. It is unimportant that good consequences might result from breaking the laws, for the laws are absolute. Therefore, any result following on their nonobservance must be considered wrong by definition.

In contrast, Joseph Fletcher, another theologian, eschews the metaphysical for a pragmatic, consequentialist approach. To Fletcher, no act is bad or good in and of itself: the important consideration is whether or not it enhances or detracts from human well-being. Fletcher argues for the careful consideration of each issue in Genetic Engineering, so that a final decision may be reached by a rational evaluation of risks and benefits, rather than by the metarational acceptance of nonuniversal moral guidelines. However, Fletcher shows a tendency to subordinate individual choices to "the good of society." For example, he would consider it irresponsible, immoral, and reprehensible to knowingly give birth to a genetically afflicted child, no matter whether the parents might prefer having such a child to not having a child at all.

The work of molecular biologist Robert Sinsheimer enables him to see clearly the vast potential for amelioration of genetic problems; perhaps this accounts for his joining Fletcher in adopting a consequentialist philosophy, although a more liberal one. He advocates that we proceed, but with proper caution. Sinsheimer believes that all human advances have occurred by trial and error, and that the first general reaction is invariably negative. He proposes the following goals:

> The aims of genetic technology should be to enhance the capaci-
> ties of each individual to comprehend and to cope freely with the ↰
> complexities of interactive society, to enlarge the internal margin of
> humanity, to transcend our conceptual limitations.[3]

Kass and Ramsey state their position strongly and unequivocally. They say that Genetic Engineering is wrong, that it will lead to evil, and that in fact, it is *itself* evil. The opposing argument is couched in less definite terms: Genetic Engineering can be beneficial, under certain conditions, and if we move with due caution. On consideration then, it becomes apparent why so many popular science writers and their readers maintain a basically negative attitude toward Genetic Engineering. A fundamentalistic assertion too often seems more compelling than an agnostic scratch of the head.

As I mentioned earlier, Genetic Engineering frequently elicits comments to the effect that its development will simultaneously lead us back to the Nazis and ahead to *Brave New World*. But this is basically a false issue. Tyrannical governments don't need genetical knowledge to subjugate populations or to perpetrate atrocities. No Genetic Engineering was used to produce the hundreds of millions of Chinese who obediently contemplate the sayings of their Chairman as they write symphonic music by committee. Misuse of genetic information would be the result, not the cause, of despotism. As the Nobel Science Laureate Dr. Joshua Lederberg wrote:

> It is indeed true that I might fear the control of my behavior
> through electrical impulses directed into my brain but . . . I do not
> accept the implantation of the electrodes except at the point of a gun:
> the gun is the problem.[4]

Another erroneous concept is that Genetic Engineering is comparable to the physical sciences whose undirected and unrestricted development gave us atomic bombs and atmospheric pollution. This is not valid: despite the frequent use of the charged word "technology" to refer to different procedures in Genetic Engineering, genetics is a biological science and a branch of medicine. A more appropriate comparison could be made to Applied Microbiology, or "Microbial Engineering." This field has given us the antibiotics which have been responsible for saving millions of lives over the past thirty years. Of course, antibiotics have also produced a few problems: for example, resistant bacterial strains, and large numbers of troubled old people. Yet on balance, I think not many of us would choose to proscribe the use of antibiotics.

In his "Notes of a Biology-Watcher" in the *New England Journal of Medicine,* Dr. Lewis Thomas commented:

> It would not surprise me to learn that there were ancient pre-fire committees, convened to argue that thumbs might be taking us too far, that we'd have been better off with simply another finger of the usual sort.[5]

This theme is expanded by geneticist Curt Stern, when he writes:

> truth and the search for it can be suppressed not only by ill-will but also by good will. In our justified fear of mankind's misuse of its powers the cry for a moratorium on research has often been sounded. Those who defend such a moratorium do not make entries on both sides of the balance sheet. From the discovery of fire to that of nuclear power, the destructive consequences have been accompanied by beneficial ones. It is the moral tragedy of man that though the extinction of one life cannot be compensated by the preservation of even many lives he cannot avoid making choices. Moreover, it is impossible to wait with new explorations until man is a more moral being. It is not likely that he will soon attain the status of an angel, either by changing his basic nature or by being born into a perfect society. Rather, for a long time the inhuman nature in each of us will have to be overcome by slow individual effort. While this is the case, can we deny man the benefits of possible new discoveries on his genetic variability because there is also the possibility of harm arising from such discoveries?[6]

Stern's points are valid. In our anxiety, we may forget about the overwhelming infant mortality and plague epidemics that were much a part of life not so many years ago. Concern for implications and outcomes is fine and appropriate, but it is possible that we are now witnessing a pathological concern for the future. Since all the ramifications of an act can never be foretold, it is possible to conclude that the only "ethical" thing to do is nothing. And it is indeed just this approach that Kass advocates.

For as many years as we can reasonably look forward to, the goals of genetics will be those of ameliorating disease. But suppose that at some distant time in the unforeseeable future, some of the prophecies of the present-day writers of undeclared science fiction are borne out? Suppose we do eventually acquire the capacity to evolve into beings somewhat different from what we now are? I for one think that that would not necessarily be bad. As individuals or as a society, I don't believe we've yet reached that happy point where we can no longer be improved upon.

Furthermore, with the passing of years, the significance of many genetical controversies may be dramatically altered. For example, Dr. Thomas' cavemen probably would have argued strongly against permitting severely nearsighted persons to reproduce. Today, however, eyeglasses have reduced this formerly lethal genetic problem to a minor nuisance.

Intertwined through the basic fabric of my entire discussion has been the question of whether or not individuals should be free to conduct their own affairs. I believe they should. As scientists gather the genetical facts, bit by bit, we should all have the opportunity to acquaint ourselves with the data, then give thorough consideration to the issues, and finally make decisions in accord with our own concepts of right and wrong. In this way, the exercising of genetical options would remain an individual prerogative.

ך

Introduction

Principles of Genetics

Before diving into the mainstream of Genetic Engineering, some readers might wish to acquaint themselves with explanations of basic genetic concepts and definitions of fundamental genetic jargon. Certainly, I would be at a severe disadvantage in trying to make sense out of a book on automotive repair unless someone were to first explain to me the component parts of the engine. Those readers whose knowledge of chromosomes exceeds my comprehension of carburetors should proceed directly to the first chapter, and perhaps rely for periodic assistance upon the glossary at the end of the book.

Genetic Terminology

Genes are the basic units of inheritance; they consist of specific lengths of interlocking double-stranded deoxyribonucleic acid (DNA). A particular gene codes for (directs) the production of a particular protein or part of a protein. Some of these proteins order embryonic development; others are *enzymes* (catalysts which control cellular metabolic reactions); still others become part of the body structure. The observable manifestations of the actions of genes are called *traits* or *characters.* A given trait may be due to the action of one gene or of many. Most genetic traits simply represent normal variations: for example, eye or hair color.

If the DNA comprising a gene should undergo a structural alteration, or *mutate,* the gene will then code for a correspondingly altered protein. Sometimes, the mutation will be *neutral:* the new protein can do its job as well as the original. Occasionally the mutation is *favorable,* in that the restructured protein is even more efficient. But most often, mutations are unfavorable, meaning that the different protein is of the wrong size or shape, so the necessary biological function cannot be performed. This compromises the health of the

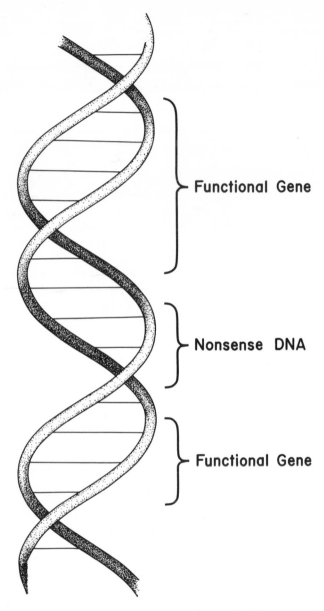

Fig. I.1: A segment of a chromosome. As described by Watson and Crick in 1952, the DNA in a chromosome exists as an interlocking double helix. Regions of functioning DNA (genes) alternate with lengths of nonsense DNA, the function of which is uncertain.

individual, as manifested by the appearance of abnormal traits, which we speak of as *genetic diseases*. Although radiation and some chemicals increase mutation rates in laboratory animals, the basic causes of gene mutation remain obscure.

The word *chromosome* literally means "colored body," and these structures were so named by Dr. W. Waldeyer of Germany in 1888, some fifteen years after they had been observed and described by Schneider, during his studies on cell division in a flatworm. Chromosomes contain proteins, but by far their major component is DNA. A widely used analogy compares the arrangement of genes on chromosomes to "beads on a chain." This has some validity, but in fact, not all the DNA of chromosomes is genetically active. Functioning genes, or *sense DNA*, alternate with areas of *nonsense DNA* (Fig. I.1). This nonsense DNA either serves some obscure function, or exists as a vestigial reminder of our genetic evolution.

Every plant and animal species has a specific number of chromosomes: in man, this is 46. Chromosomes (and therefore genes, as well) occur in pairs, one member of each pair having been inherited from the father and one from the mother. The members of a chromosome pair are called *homologous chromosomes*, and the two members of any gene pair are called *alleles*.

Chromosomes can be visualized with ordinary light microscopes, but individual genes cannot. Therefore, the functional competence of a particular gene can only be inferred indirectly by study of the trait it produces. No one knows how many genes are located on our 46 chromosomes, but estimates range from 50,000 to upwards of a million. Chromosomes reside in the nuclei of cells, and one of the most fascinating questions in genetics relates to the fact that all nucleated body cells contain exactly the same 46 chromosomes and exactly the same gene complement. This being so, it would be most useful to know what is responsible for the differential activation of genes, so that one group of embryonic cells is induced to form a liver and do the work of a liver, while another group of cells becomes a heart.

Chromosomes are visualized and studied by bursting open the cells that contain them, so that the chromosome groups spread out onto the surface of a glass slide. These groups are photographed; then, the individual chromosomes in each group are cut out, paired, and arranged according to their size (large, medium, small) and to the position of the constricted region, or *centromere* (central, off-center, or near the end). This procedure is called *karyotyping*, and a *karyotype* is a photograph of the final, standardized arrangement of the chromosomes of a cell (Fig. I.2).

Until 1970, chromosomes could only be separated into groups of similar appearance, labeled A through G. However, it was then discovered that either specific variations in the customary staining technique, or the use of new fluorescent dyes would produce patterns of alternating dark-staining and light-staining bands specific for each chromosome pair (Fig. I.3). These staining characteristics do not relate to the presence of specific individual genes, but to variations in the basic chemical composition of different regions of DNA.

⟨Genetically speaking, the only difference between a man and a woman is one chromosome. Male-determining genes are located on the little Y chromosome; embryos without a Y differentiate as females. Thus, the normal *sex-chromosome* complement of males is XY, that of females, XX. The other 22 pairs of chromosomes, identical in both sexes, are called *autosomes*⟩

The question sometimes arises as to why the difference in sex chromosomes does not create a genetic imbalance between men and women. This is answered by two facts. First, the Y chromosome does not seem to carry genes for any function other than sex determination; second, early in the embryonic life of females, one of the two X chromosomes in each cell enters a state of functional inactivation. The result is that both men and women are left with 45 identical chromosomes which carry the genes for general body functions.

The inactivation of the second X chromosome is the basis for the

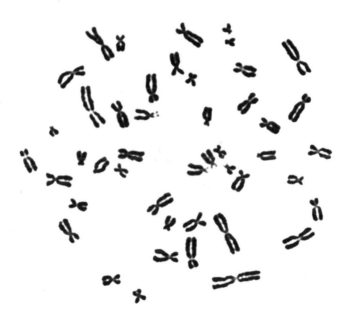

Fig. I.2:   (a) A chromosome group from a burst cell, as it appears under the microscope.

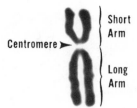

Centromere ► { Short Arm { Long Arm

46, XX

A B

C

D E

F G X X

46, XY

A B

C

D E

F G Y X

(b) Human karyotypes. Note that the only chromosomal difference between males and females is in the second sex chromosome.

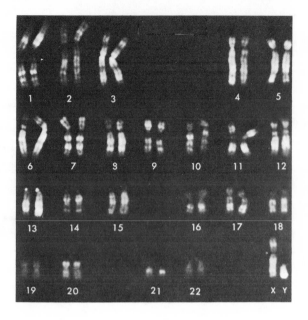

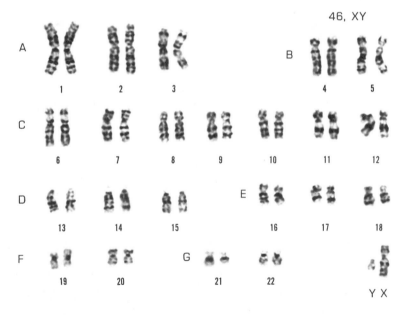

Fig. I.3: Human karyotypes, using the new fluorescent and modified conventional staining techniques to produce a banding pattern characteristic for each chromosome pair. (Fluorescent karyotype courtesy of Dr. Irene Uchida, McMaster University, Hamilton, Ontario.)

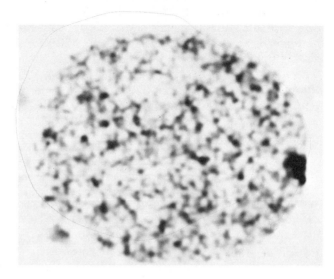

Fig. I.4: Female sex chromatin (Barr body), present in cells from chromosomally normal women, but not men.

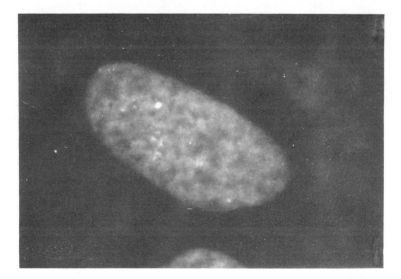

Fig. I.5: Male sex chromatin ("flashing Y"), present in cells from chromosomally normal men, but not women.

technique used to rapidly screen for a person's chromosomal sex. Largely due to the work of Dr. Murray Barr in Canada, it is known that when cells from a woman's cheek lining are placed on a slide and treated with conventional cell stains, a dark spot appears on the inner aspect of the nuclear membrane. This spot is called the *female sex chromatin,* or *Barr body* (Fig. I.4). Dr. Mary Lyon of England has shown that the Barr body is in fact the inactivated X chromosome. There are also *male sex chromatin:* since the Y is the most highly fluorescent chromosome, treatment of a man's cheek cells with fluorescent stain will produce a bright spot within the nucleus (Fig. I.5).

Many terms are used to describe various characteristics of chromosomes and genes. *Ploidy* refers to the number of chromosomes possessed by a cell or an individual. Any cell with the proper number of chromosomes is termed *euploid,* while cells with abnormal chromosome complements (for example, 45 or 47) are *aneuploid.* Human body cells normally have two complete sets of 23 chromosomes: they are *diploid.* Some liver cells, however, are tetraploid (92 chromosomes), as are some cells of the amniotic membrane. However, normal human cells never seem to be *triploid* (69 chromosomes).

Recalling that genes occur in allelic pairs, we can consider the concept of *zygosity.* If both members of any particular pair of alleles are identical (for example, if both genes code for blue eyes), the person is said to be homozygous for those genes. But an individual with one gene for blue eyes and one for brown eyes is *heterozygous* for eye-color genes. However, since men have only one X chromosome, they are *hemizygous* for all genes carried on the X. It should be remembered, too, that since women randomly inactivate one X chromosome in every cell, each female cell is *functionally hemizygous.*

When a person is heterozygous for a particular allelic gene pair, one gene usually expresses its trait while its partner remains "silent." For example, someone with one gene for brown eyes and one for blue eyes usually will have brown eyes. In this hypothetical case, the gene for brown eyes is said to be the *dominant* allele, and the gene for blue eyes *recessive.* Recessive genes will be expressed only when their bearer is homozygous for them, or if they are carried on the X chromosome of a male. A few genes are *co-dominant:* a person will be of blood type AB if he has one A gene and one B gene. Lest anyone fear that his body is nothing more than a gigantic molecular rat race, it can be explained that recessive genes usually are of this character because they are "negative" genes. For example, all babies initially have blue eyes; those who later develop brown eyes do so because their genes direct the deposition of brown pigment in the iris. This recessive condition of "occupying a space but failing to produce anything" is the basis of many genetic diseases.

The replication of chromosomes and their transmission from cell to cell are important processes. You may have wondered how it is that men and women manage to pass on only half of their chromosomes to their offspring. If there were no mechanism to accomplish this, the chromosome number would double with each generation, and before too long, our bodies would be nothing but writhing masses of billions of chromosomes. The generational halving of chromosome number is accomplished by *meiosis,* or *reduction division.* In the *germ cells* (reproductive cells) of the ovary or testis, each chromosome lines up opposite its homologous partner, and in the ensuing production of eggs or spermatozoa, one member of each chromosome pair is cast out of the cell. Thus, each mature egg or sperm cell is *haploid,* containing, in the human, one set of 23 chromosomes (Fig. I.6). Both eggs and sperm contain 22 autosomes, and of course each egg has an X. Roughly half of a man's sperm contain X chromosomes, and the other half have Y's; this results from the pairing of the X with the Y at the onset of meiosis. Thus, the sex of the offspring is determined by the sex chromosome in the fertilizing sperm.

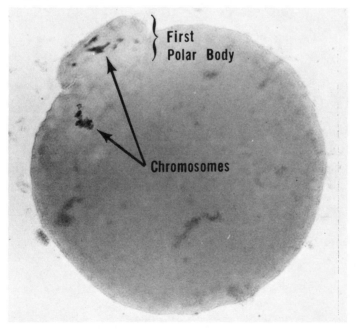

Fig. I.6: In this mouse egg, meiotic division has resulted in the retention of half the chromosomes, while the other half have been cast off into the polar body, which will degenerate. The number of chromosomes in eggs can be verified as haploid by bursting the egg cells on a slide, thus permitting karyotyping.

Fig. I.7:   The beginning of a new organism, a mouse. This animal has 40
            chromosomes. The entire process of fertilization and cell division is
            identical in man.

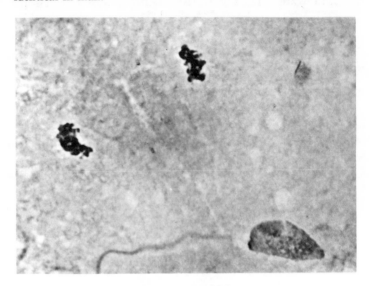

(a) Fertilization. The sperm is seen within the egg.

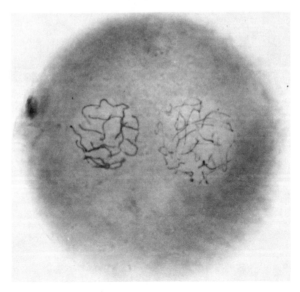

(b) The two chromosome groups of 20 move together. The larger
group is from the sperm.

(c) Diagrammatic representation of chromosome replication, achieved by progressive "unzipping" of the double helix. Each chromosome in Fig. 7b is undergoing this process.

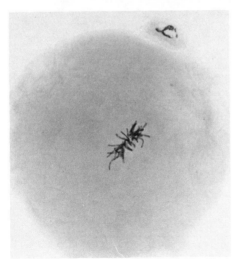

(d) The two groups of chromosomes, now replicated, merge and line up to undergo the first mitotic division.

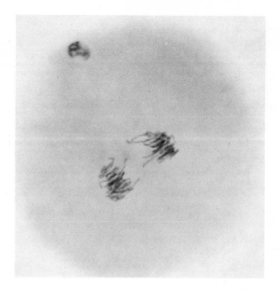

(e) The 40 replicated chromosomes have divided and have begun to separate into two equal groups.

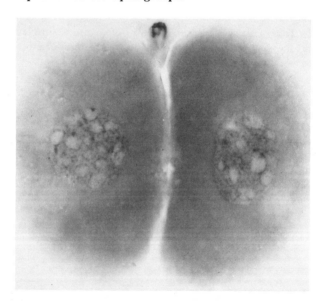

(f) The cell has divided, producing a two-cell mouse embryo. Each nucleus contains 40 chromosomes, but at this stage of the cell life cycle, the individual chromosomes cannot be seen. (Photographs in Figs. 7a, b, d, e, f. from R. P. Donahue, Fertilization of the mouse oocyte: Sequence and timing of nuclear progression to the two-cell stage. *J. Exp. Zool.* **180 (1972): 305-318.)**

After the sperm penetrates the egg, the two haploid chromosome groups move together, and gradually merge (Fig. I.7). Around this time, two important events occur. First, the DNA of the chromosomes replicates itself: the double strands of DNA in each chromosome progressively separate, rather like the opening of a zipper, and then each strand serves as a template onto which its new complementary strand is laid down. Thus, all the newly produced chromosomes are composed of one "old" strand and one "new" strand of DNA.

Next, the "double chromosomes" line up at the middle of the cell and divide in concert, so that an equal number of "single chromosomes" travel to opposite ends of the cells. Finally, the cell itself divides along its midline. Thus, the net result is a two-cell embryo with the normal number of chromosomes in each cell.

This process of chromosome replication and separation, followed by cell division, is called *mitosis*. By repeated mitoses, the one original cell will give rise to all the cells in the body of the individual. Mitotic cell division goes on throughout our lives, continually replacing those cells that have lived their allotted spans.

I'll conclude this section of the chapter by defining a few terms that often confuse people. A *genetic disease* is an illness caused by malfunction of a single gene or of many genes. *Familial disease* frequently is used synonymously, and sometimes this is accurate. But strictly speaking, familial diseases are those that are passed from generation to generation, diseases which "occur in families." For example, both a mother and her child may suffer from tuberculosis; this situation represents a disease that is familial, but not genetic. Conversely, if a gene mutation occurs in a one-cell embryo, and causes such severe derangement that the fetus is either aborted or dies shortly after birth, this would represent a genetic, but not a familial disease.

Congenital diseases are illnesses that are present at birth. A baby infected as an embryo with German measles (rubella) virus may be born with nongenetic congenital disease. Huntington's chorea is genetically determined, but does not appear until adult life; therefore, it is not congenital. Down's syndrome (mongolism), however, is both congenital and genetic.

Genetic fitness refers not to the "goodness" or "badness" of genes, or to the presence or absence of genetic diseases, but to a person's ability to pass on his genes to the next generation. Hemophiliacs are less fit than the general population, because fewer of them marry and have children. However, thanks to new methods of treatment of the disease, the fitness of hemophiliacs is higher now than formerly, though the gene itself has not changed. A fertile hemophiliac is more fit than a football player who is sterile due to an injury to his testicles. The fitness of the football player is, in fact, zero.

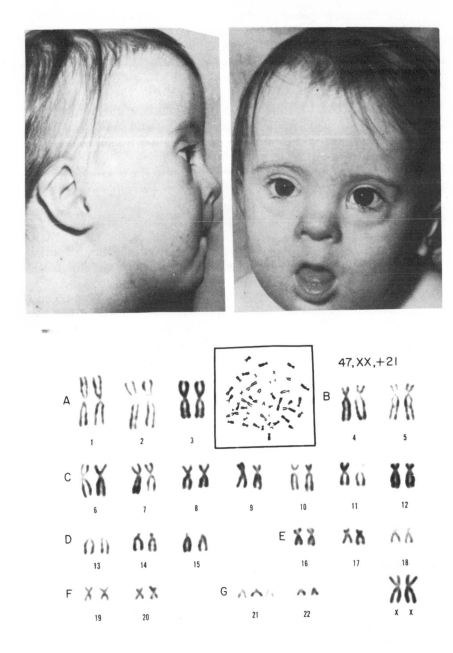

Fig. I.8: The appearance and karyotype of a child with Down's syndrome, due to trisomy-21. (From D. W. Smith, *Recognizable Patterns of Human Malformation.* Philadelphia: W. B. Saunders Co., 1970.)

Mechanisms of Inheritance of Genetic Diseases

Genetic diseases can be discussed under the following headings: chromosomal abnormalities; single gene defects, both dominant and recessive; and multifactorial diseases.

Chromosomal Abnormalities

Chromosomal aberrations are basically of two types: abnormalities of number and abnormalities of structure. Since genes are arranged in a linear fashion along the chromosomes, the fundamental principle can be set forth that any numerical or structural chromosomal anomaly that results in an individual's having too many or too few genes will produce physical or mental defects. Although structural or numerical chromosomal errors occur once in every 200 live births, they are not inevitably associated with physical or mental defects. This variability probably depends upon whether the chromosomal anomaly involves functional genes or only regions of "nonsense" DNA.

Errors of chromosome number. Approximately 40 percent of miscarriages in the first three months of pregnancy are associated with aneuploidy, an abnormal chromosome number. Since 15 percent of recognized pregnancies are miscarried, this means that at least 6 percent of all conceptions result in aneuploid embryos.

Any numerical chromosomal (and therefore genetic) imbalance puts an embryo at a severe disadvantage, and most aneuploid states produce such severe developmental abnormalities that the embryos almost invariably succumb very early. But a few chromosomal anomalies produce less severe effects, so that a proportion of the affected fetuses reach term and are born alive. The best known of these conditions is Down's syndrome, or mongolism, caused by *trisomy* ("three chromosomes") for chromosome number 21. Others are trisomy-18, trisomy-13, trisomy-8, trisomy-X, XXY (Klinefelter's syndrome), XYY, and monosomy-X.

The incidence rates of all these conditions except the last two are strikingly associated with increased maternal age (Table I-1). By the time she is forty, a woman has almost a 3 percent chance of giving birth to an aneuploid child; approximately half this risk is for Down's syndrome.

Down's syndrome was first described clinically in 1866, but it was not until 1959 that the French cytogeneticists Lejeune and Turpin discovered it to be caused by an extra chromosome number 21 (Fig. I.8). The patients usually can be identified by their characteristic facial appearance; in addition, they are short of stature, have abnormal hand and fingerprint patterns, and have weak muscle tone. One-third have serious heart defects. The IQ is usually between 20 and 50. About one-

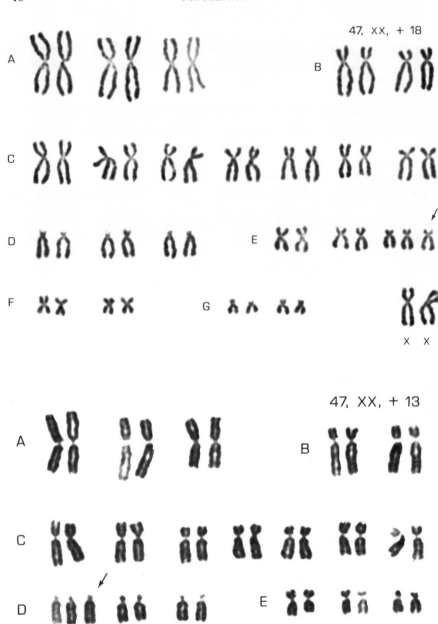

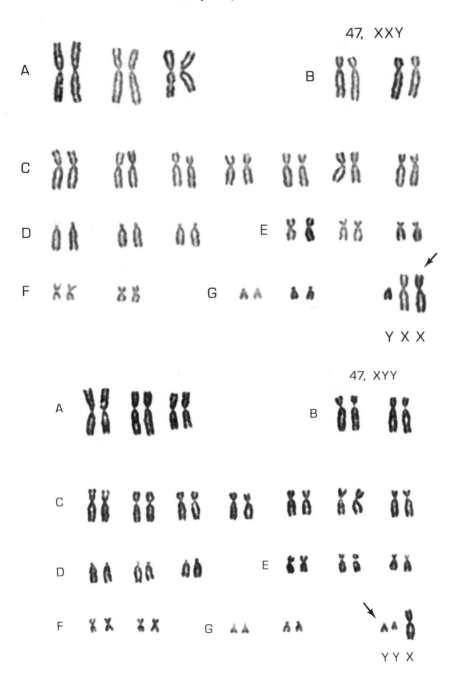

Fig. I.9: The abnormal karyotypes of trisomy-8, trisomy-13, and the Klinefelter and **"XYY"** syndromes.

Table I-1

Risk of Babies With Down's Syndrome and Other Aneuploidies Related to Maternal Age

Mother's Age at Birth of Child	Risk of Down's Syndrome	Risk of Any Aneuploidy
Under 30	1:1500	1:750
30-34	1:1000	1:500
35-39	1:300	1:150
40-44	1:100	1:50
45 and over	1:40	1:20

third of these patients live to adulthood, one-third die before the age of ten, and another one-third die during adolescence. However, with improving medical and surgical care there is a constantly improving survival potential for patients with Down's Syndrome.

Patients with trisomy-18 are usually female, small, and spastic. They have characteristic hand and foot deformities and severe anomalies of the internal organs. Most are stillborn or die shortly after birth: only about 10 percent survive to the age of two. Mental deficiency is profound (Fig. I.9).

The features of trisomy-13 include central facial defects such as cleft palate, fused or malformed eyes, and deformed nose. In addition, there are major visceral malformations, and severe maldevelopment of the brain with mental retardation. About 5 percent of such children survive the newborn period (Fig. I.9).

Aside from mental retardation, patients with trisomy-8 present with small chins, short necks, and deformities of the ribs and vertebrae. This condition has only recently been described, and the long-term survival potential has yet to be defined.

Anomalies of the sex chromosomes are less serious than autosomal aberrations. This probably relates both to the cellular inactivation of all X chromosomes but one, and to the fact that no genes other than those for male sex determination have been localized to the Y. Therefore, developmental errors caused by improper numbers of active genes would be less likely to occur when the sex chromosomes, rather than the autosomes, are involved.

Another general rule is that a deficiency of genetic material is more harmful than an excess. Monosomy-X is seen in 49 miscarriages for every live-born case. In addition, autosomal monosomies are extremely rare, even in miscarriages; therefore, it's assumed that such conceptuses never even develop far enough to be recognized as pregnancies.

Patients with monosomy-X are female, and have the well-defined Turner syndrome, which consists of short stature, webbed neck, and absence of sexual maturation. This latter characteristic is related to the failure of the ovaries to develop and then produce sex hormones and release eggs. Administration of estrogens to these patients will result in both sexual maturation and the capacity for normal sex relationships, but the infertility is not amenable to treatment. A minority of patients manifest heart abnormalities or high blood pressure. Although many show difficulty in dealing with spatial or geometric concepts, there is no more mental retardation among these patients than in the general population.

Most patients with trisomy-X seem to be entirely normal females, but a few show mild mental deficiency, slight speech or motor deficits, or difficulty in relating to people. There is also some suggestion that they may be at increased risk for menstrual disturbances.

Klinefelter's syndrome is caused by an XXY chromosome makeup (Fig. I.9). In three-quarters of such cases, the extra X arises in the egg; in one-quarter, in the sperm. These patients are tall, eunuchoid males who are infertile due to testicular maldevelopment. One-fifth grow breasts, which must be removed surgically. Mental retardation is not the rule, but a mild to moderate deficit occurs in a significant minority of patients. Psychopathology is also a frequent problem, consisting of excessive passivity, difficulty with interpersonal relationships, and failure to cope with situational demands.

About one in every thirty inmates of mental or penal institutions has 47 chromosomes, with either an XXY or an XYY sex-chromosomal constitution. The latter condition obviously is always of paternal origin (Fig. I.9). Since it was first recognized in maximal-security prisoners, it was widely publicized as "the criminal karyotype." It has since become apparent, however, that even though XYY individuals are at some increased risk for mild mental retardation, psychopathology, and a tendency to impulsivity and lessened restraint, the large majority of XYY men are perfectly normal, law-abiding citizens. I would emphasize that there are no known "criminal chromosomes" or "criminal genes." The unfortunate publicity relating to XYY has led some geneticists to fear unnecessary stigmatization, and so not to tell the parents when an infant or a child is found to be carrying this chromosomal anomaly. While I can understand the rationale of this policy, I still think it constitutes unethical withholding of information, and ought not to be used as an alternative to extensive, sympathetic counseling and close supportive follow-up.

Errors of chromosome structure. During replication of the chromosomes in the sperm or the egg, several anomalies can arise. Gene *duplication* may occur, and conversely, a chromosomal fragment may be lost, causing gene *deletion* (Fig. I.10). Either situation can result in a defective embryo. In addition, *translocations* sometimes arise, involving the attachment of a whole chromosome or a fragment thereof to another chromosome. When no net gain or loss of genetic

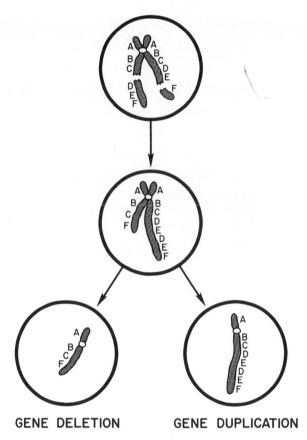

GENE DELETION   GENE DUPLICATION

Fig. I.10: A possible mechanism whereby gene duplication or deletion could arise, between the times of chromosomal replication and division. There is seen to be transitory chromosome breakage, followed by faulty repair.

material occurs, the translocation is said to be *balanced,* and the translocation carrier will be normal. However, unbalanced translocations usually are found in defective individuals (Fig. I.11).

The healthy carriers of balanced translocations are at risk for unbalanced offspring. About 2 percent of all cases of Down's syndrome are caused by a translocation involving the attachment of a third number 21 chromosome to a D- or a G-group chromosome. Half of these unbalanced translocations arise as new events in the embryos, but the other half are transmitted from balanced carrier parents, as illustrated in Fig. I.12. This latter situation accounts for the rare cases of Down's syndrome that "run in families." Either parent can be the

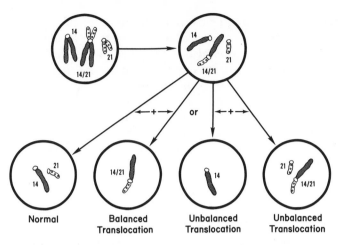

Fig. I.11: Translocations. These consist of rearrangements of genetic material between two different chromosomes. A balanced translocation is a chromosomal rearrangement where there is no net loss or gain of genes. The genetic loss or gain in an unbalanced translocation usually produces genetic disease.

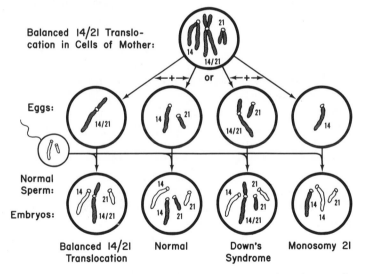

Fig. I.12: The possible offspring of a 14/21 balanced translocation carrier. The embryo with monosomy-21, at the far right, is not viable, and would be miscarried early in pregnancy. Because unbalanced eggs and sperm are at a general disadvantage in fertilization, and because unbalanced embryos are more likely to be miscarried, the actual risk of Down's syndrome in a live-born child is less than the theoretical 33 percent: it is 15 percent when the mother is the carrier and 5 percent when the father is the carrier.

carrier, but more frequently it is the mother, probably because an abnormal sperm is at greater disadvantage than an abnormal egg.

Causes of chromosomal abnormalities. We really know little of the causes of numerical and structural chromosomal anomalies. This field is currently the subject of considerable research.

Perhaps the most striking correlation is that of increased maternal age with trisomies. Most geneticists believe that this reflects the fact that all of a woman's eggs are formed while she herself is a fetus. Thus, the ova she releases near the end of her reproductive life are themselves more than 40 years old. In contrast, during a man's reproductive years, spermatozoa are continually being formed.

The process by which an extra chromosome becomes included in the egg is called *nondisjunction,* because it involves failure of a homologous pair of chromosomes to separate (disjoin) during meiosis, so that they both end up in the egg, rather than one being cast off into the polar body (Fig. I.13).

It has been postulated that one cause of nondisjunction may be a genetic abnormality. In plants and in some invertebrates, there exist genes that control normal meiotic chromosomal pairing and separation. Malfunction of hypothetical similar genes in humans has been put forth as one possible explanation both for the existence of families with multiple trisomic individuals far in excess of the number expected by chance, and for the fact that women who have had one trisomic child are at increased risk to have another.

Prepregnancy maternal radiation seems to have a small but significant association with aneuploid offspring, both in humans and in experimental animals. Furthermore, exposure of cells in laboratory vessels to ionizing radiation, chemicals, or viruses will produce breakage and fragmentation of chromosomes. Such damage might lead to translocations, or to deletion or duplication errors in ensuing replication. It's also possible that such structurally abnormal chromosomes might be more likely to undergo nondisjunctional errors.

Unfortunately, we have no knowledge whatever regarding the ultimate cause(s) of chromosomal aberrations. Until we discover the basic cellular biochemical processes whose derangements result in abnormalities of chromosome number and structure, it's not likely that we'll be able to accomplish much in the way of prevention.

Single Gene Defects

Diseases caused by mutation of either a single gene or a pair of alleles are usually divided according to the character of the offending gene: autosomal dominant, autosomal recessive, or X (sex)-linked.

Autosomal dominant diseases. These are illnesses caused by a mutation in a single autosomal gene. A mutant dominant gene is

Table I-2

Some Common Autosomal Dominant Diseases

Achondroplastic Dwarfism
Ehlers-Danlos Syndrome
   ("India Rubber Men")
Huntington's Chorea
Marfan Syndrome
Myotonic Dystrophy
Neurofibromatosis
Polycystic Kidneys
Porphyria
Retinoblastoma

usually a "positive abnormal gene," in that it produces structural developmental abnormalities which cannot be masked by a normal allele. Mental retardation is not a conspicuous feature of most autosomal dominant diseases. Some of the more common ones are listed in Table I-2.

Depending upon the specific disorder, a variable proportion of patients with autosomal dominant diseases inherit their mutant genes, while others have unaffected parents, and so represent new mutational events. Women who are made unhappy by the association of chromosomal errors with advanced maternal age may take some comfort from the fact that many fresh autosomal dominant mutations seem to be associated with advanced age of the father.

Since only one of every pair of homologous chromosomes is included in any one particular egg or sperm, a person affected with an autosomal dominant disease transmits the trait to half his children — the half which receive the chromosome with the mutant gene. The other half of the children receive the chromosome with the normal gene (Fig. I.14). Since this is so, unaffected relatives of patients with autosomal dominant diseases are at no risk for having affected children. And since the gene in question is located on an autosome, males and females are affected equally.

A point deserving of strong emphasis is the fact that the risk of affected offspring is 50 percent *for each pregnancy* of an affected parent. There is a common misconception to the effect that as the number of affected children increases, the chance of further affected offspring progressively diminishes. In truth, however, the inclusion of chromosomes in each egg or sperm is an independent event. "Chance has no memory," and the likelihood that an eleventh child will be affected is the same whether the first ten children were all affected or all unaffected.

Autosomal recessive diseases. In contrast to dominant diseases, recessive conditions usually result from genes that mutate into forms

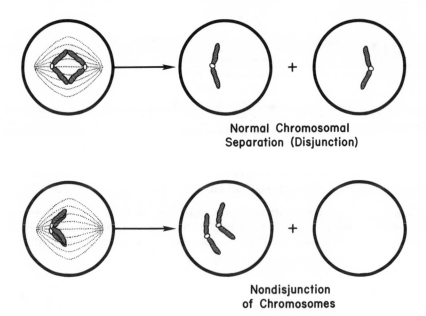

Fig. I.13: Diagrammatic representation of nondisjunction: immediately prior to meiotic reductional division, the chromosomes of a particular pair line up together, rather than opposite each other.

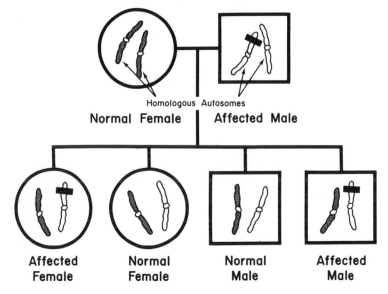

Fig. I.14: Genetic mechanics of autosomal dominant inheritance. By genetic convention, females are denoted by circles, males by squares. The black cross-bar denotes the mutant gene.

which have "negative" effects. Recessive gene mutations occur as frequently as dominant ones, but clinical disease is not the immediate result, because the normal allele compensates for the functional absence of its partner. The person carrying such a recessive mutation is called an *asymptomatic carrier.*

Autosomal recessive disease occurs only when there is a mating between two heterozygous carriers, a situation where the offspring may receive two mutant, nonfunctional genes. As illustrated, this will happen in one of every four children of carrier parents (Fig. I.15). Of the other three children, one will receive the two normal genes, and the other two will each receive one normal and one mutated gene. These two new carriers will themselves have affected children only if they marry other carriers for the same mutation. And since the incidence of most gene mutations in the general population is low (usually under one percent), it becomes apparent why autosomal recessive disorders characteristically appear in siblings, but not in successive generations.

It's estimated that all of us carry 3 to 5 mutant genes, which only cause trouble when we are unlucky enough to have children by a carrier for a matching mutation. Since family members have genes in common, a marriage between close relatives is more likely to effect a pairing of similar mutant genes. Thus, autosomal recessive diseases are more common in *consanguineous* matings.

A large proportion of autosomal recessive disorders are called *inborn errors of metabolism,* because they are caused by deficiencies of one or another of the multitudinous enzymes that act as catalysts in intracellular metabolic reactions. In the mutation homozygote, clinical disease results either from deficiency of the end product of the metabolic reaction or from the cellular accumulation to toxic levels of an intermediate substance that normally should be metabolized by the missing enzyme. A carrier is unaffected because his single normal gene can by itself produce enough enzyme to sustain cellular function. Inborn errors of metabolism may be subdivided according to the class of metabolite which cannot be degraded; some of them are listed in Table I-3.

X-linked diseases. These conditions are sometimes called "sex-linked," but the term "X-linked" is currently more in favor as being more specific, since only genes for male sex determination have been localized to the Y chromosome. There are a few X-linked conditions that are inherited in a dominant fashion, but most X-linked genes seem to behave in a recessive manner. By far, the most significant X-linked recessive diseases are hemophilia and Duchenne muscular dystrophy (Table I-4).

The reason for considering X-linked recessive conditions separately from autosomal recessives lies in the fact that males are hemizy-

Table I-3

Some Common Autosomal Recessive Diseases

Inborn Errors of Metabolism
    Errors of Carbohydrate Metabolism: Galactosemia
                                     Glycogen Storage Diseases
    Errors of Amino Acid Metabolism: Phenylketonuria
    Errors of Lipid Metabolism: Gaucher's Disease
                         Niemann-Pick Disease
                         Tay-Sachs Disease
    Errors of Mucopolysaccharide Metabolism: Cystic Fibrosis
                                    Hurler Syndrome
                                    (Childhood "Gargoylism")
Miscellaneous: Albinism
                   Congenital Adrenal Hyperplasia
Other Autosomal Recessive Disorders
    Sickle Cell Anemia
    Spino-cerebellar Ataxias

gous for the X chromosome. Therefore, in males, a single dose of a "negative" recessive gene is sufficient to produce disease, because there can be no normal allele to compensate. In females, however, the normal allele on the second X chromosome prevents clinical disease, so the distaff possessor of a deleterious X-linked gene is a carrier, much as though the gene were located on an autosome. There is one relatively minor difference, and this is discussed in the chapter on Genetic Screening.

Table I-4

Some Common X-linked Recessive Diseases

Color Blindness
Hemophilia
Lesch-Nyhan Syndrome
Muscular Dystrophy

As shown in Fig. I.16, a carrier woman passes on her defective gene to half her children, just as any gene carrier does. Since all her daughters receive a normal allele from the father, half the daughters are genetically normal and the other half, like the mother, are carriers. But since all the sons receive a Y from the father — that's why they're sons — the half that get the abnormal gene from the mother will have the clinical disease. The sons that are lucky enough to get the mother's normal X are unaffected.

On the other hand, when a hemophiliac man reproduces, all his daughters will be carriers, while all his sons (who receive his Y, not his X) will be unaffected (Fig. I.17). Theoretically, a male hemophiliac

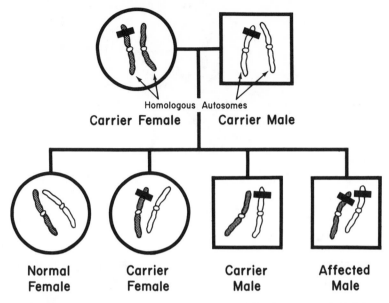

Fig. I.15: Genetic mechanics of autosomal recessive inheritance, Note that two carriers for the same mutation have a 25 percent chance of having a genetically normal child, a 50 percent chance of having a carrier child, and a 25 percent chance of having a clinically affected child.

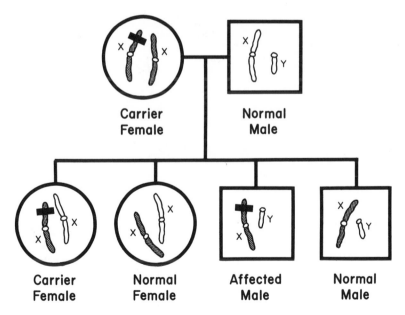

Fig. I.16: Genetic mechanics of X-linked recessive inheritance, through a carrier female.

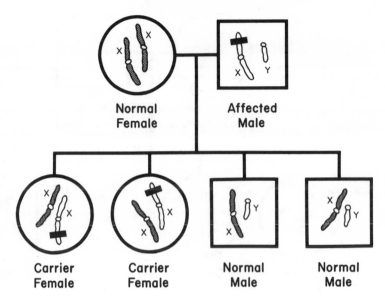

Fig. I.17: **Genetic mechanics of X-linked recessive inheritance, through an affected male.**

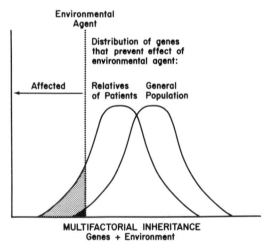

Fig. I.18: **Presumed genetic-environmental interaction to produce cleft palate, a multifactorial birth defect. A small number of people in the general population have too few genes to counteract the effect of hypothetical environmental factors that tend to produce cleft palate (right distribution curve). But since family members have large numbers of genes in common, the close relatives of patients with cleft palate constitute a high-risk subpopulation for low numbers of "anti-cleft palate genes" (left distribution curve), and so are at higher risk for offspring with this disease.**

could father a hemophiliac daughter, but to do so, his wife would have to be a carrier. This is an exceedingly rare situation.

Thus, in studying pedigrees of families in which an X-linked trait occurs, one sees characteristic features. Only males are affected by disease, and an affected man who reproduces never passes the affliction to his sons. However, since his daughters are carriers, he may have affected grandsons: the relationship between affected males can be traced through carrier females.

Multifactorial Conditions

Many human traits are determined not by a single pair of allelic genes, but by multiple gene pairs: they are *polygenically* determined. More properly, such conditions are called *multifactorial traits*, because they result ultimately from the interaction of the multiple gene pairs with environmental factors.

Some normal human characteristics such as intelligence and height are determined by multifactorial mechanisms. Basically, the genes delineate a range of potential: the more "smartness" genes or the more "tallness" genes a person has, the more intelligent or the taller he is capable of becoming. But this range of capacity is influenced by such environmental factors as the type of education and the amount of encouragement received, or the quantity and quality of food consumed. Polygenes for different traits are distributed in the population according to the well-known normal distribution, or bell-shaped curve. Most people possess an average number of polygenes for a given trait, while a few have high or low numbers.

Thus, it can be appreciated that some cases of mild to moderate mental retardation are of a multifactorial origin. The retardates are those individuals at the low end of the curve for "smartness" polygenes, who in addition received poor intellectual stimulation. Polygenic inheritance is also thought to be responsible for the observed increase in mental retardation in consanguineous matings. Since related people have larger numbers of genes in common, it's assumed that consanguinity increases the chance in the offspring for interaction of similar "negative intelligence" genes.

Many common birth defects are multifactorial; some are listed in Table I-5. These are thought to be caused by a polygenically determined heightened susceptibility of some individuals to a particular environmental factor that tends to interfere with normal fetal development. These environmental agents are in all cases so far only hypothetical.

Using cleft lip and palate as an example, suppose an agent exists which tends to impair normal palatal closure in the embryo. The few persons in the general population with small numbers of polygenes for counteracting this factor would be born with cleft palates (Fig. I.18).

Table I-5

Some Common Multifactorial Diseases

Anencephaly/Spina Bifida
Cleft Lip and Palate
Club Foot
Congenital Heart Diseases
Congenital Hip Dislocation
Diabetes Mellitus
Epilepsy

Relatives of these patients would be expected to have fewer "anti-cleft palate" genes than the members of the general population, and so would be at higher risk for clinical disease. The closer the relationship to a patient, the more genes in common, and therefore, the greater the risk of disease. The population incidence of cleft lip and/or palate is about one in a thousand newborn babies, but the risk for a couple's subsequent children after the birth of one affected child is about 4 percent. This is an *empiric risk*, based on statistical study of large populations, rather than on knowledge of precise genetic mechanisms of transmission, as is possible with single-gene traits. Also unlike monogenic inheritance, where recurrence risks remain constant in spite of the number of affected family members, the recurrence risk for multifactorial diseases increases progressively with the birth of each new affected child. Presumably, this reflects the fact that a progressively greater number of adverse polygenes possessed by a couple for a particular trait will lead to a correspondingly increasing likelihood of clinical disease in the offspring. Most multifactorial diseases have a recurrence rate of 2 to 5 percent after the first affected child. This rises to 8 to 10 percent after the second, and perhaps to 15 percent after the third.

In closing this chapter, I want to reiterate that it was not meant to be an exhaustive description of genetic principles, but rather a selective series of brief explanations to enable the reader to better comprehend the upcoming facts concerning Genetic Engineering. For those interested in delving deeper into genetic theory, some excellent texts on human genetics are listed in the bibliography.

In many places, expediency has necessitated the use of generalizations. Therefore, a special word of caution to those who might consider using the information in this chapter to try to deal with genetic problems in their own families. A far wiser course would include study of one or two of the genetics texts, followed by a scheduled visit to the nearest genetics center.

Part I

Genetics in Diagnosis and Therapy of Disease

1

Genetic Counseling

In 1902, the famous British physician Sir Archibald Garrod expressed the opinion that some rare human diseases such as albinism and alkaptonuria might be caused by genetic abnormalities. For such illnesses, Garrod coined the term "inborn errors of metabolism."

There was no rush to endorse Garrod's revolutionary effort to medicalize classical "pea genetics." In 1951, Dr. Sheldon Leonard, who first used the term "genetic counseling," estimated that there were ten genetic counselors in the United States. Twenty-four years later, there were almost 300 counseling *centers* in our country, most of them employing more than one counselor.

This proliferation of genetic counseling personnel and facilities is highly appropriate. An estimated 25 percent of all diseases are caused totally or in part by genetic malfunctions. At least one in ten patients admitted to pediatric hospitals suffers from a disease with a genetic component. But the public has found it difficult to keep pace with the rapid developments in human medical genetics. During the past year, a survey of a *highly educated* urban subpopulation revealed that only 58 percent could correctly define a gene, and only 48 percent thought that Down's syndrome was a genetic disease.

Unfortunately, the situation is not a whole lot better with regard to medical professionals. The great increase in genetic knowledge over the past decade has caught most doctors and nurses in a bind. Only recently has genetics been included in the curricula of the vast majority of medical schools. Because of this, relatively few physicians are capable of providing adequate genetic counseling, a service which involves diagnosing genetic diseases and explaining their nature and implications. Accurate and clear genetic counseling is the necessary initial step in the prevention and treatment of genetic disorders. As we shall see later in this chapter and in chapters 3 and 5, unfamiliarity with the disease entities under question can lead to terrible tragedies, through erroneous diagnoses or inaccurate explanations of modes of inheritance and recurrence risks.

Genetic Counseling Defined

The following definition of genetic counseling was drafted by the participants in a 1972 Workshop on Genetic Counseling sponsored by the National Genetics Foundation:

> Genetic counseling is a communication process which deals with the human problems associated with the occurrence, or the risk of occurrence, of a genetic disorder in a family. This process involves an attempt by one or more appropriately trained persons to help the individual or family (1) comprehend the medical facts, including the diagnosis, the probable course of the disorder, and the available management; (2) appreciate the way heredity contributes to the disorder, and the risk of recurrence in specified relatives; (3) understand the options for dealing with the risk of recurrence; (4) choose the course of action which seems appropriate to them in view of their risk and their family goals and act in accordance with that decision; and (5) make the best possible adjustment to the disorder in an affected family member and/or to the risk of recurrence of that disorder.[1]

Although this definition does not state that diagnosis is a part of genetic counseling, I would like to emphasize that ascertainment of the correct genetic diagnosis is an integral component of the counseling process.

In meeting the growing demand for counseling services, geneticists are now trying to determine how best to tailor the work to the situation. Most counseling is done in regularly scheduled university-hospital genetics clinics, but as doctors in other branches of medicine gradually become more genetically alert, geneticists are more frequently being called to bedsides. In fact, weekly in-patient genetic-counseling rounds have been established in many hospitals. Furthermore, rather than insist that all patients must necessarily come to them, some genetics teams like the ones at Jefferson Medical College, the University of Colorado, and the University of California at San Francisco have formed "flying squads," visiting different surrounding towns and cities on a revolving basis.

The Genetic Counseling Process

Who Does Genetic Counseling?

As defined, genetic counseling is done by "appropriately trained individuals." The meaning of "appropriate training," however, has not been strictly defined, and is presently the subject of considerable debate among geneticists. The general consensus is that a trainee should spend a variable length of time (perhaps two years) with a genetics clinic, and that persons without an M.D. degree should not counsel independently.

Since the range of genetic diseases is broad, the best counseling arrangement probably is a group composed of specialists in internal medicine, pediatrics, obstetrics, and laboratory medicine, all of whom have had further training in genetics. They should have access to the services of other competent specialists, especially neurologists, psychiatrists, and ophthalmologists. Genetically-oriented social workers can also be of help in dealing with many of the counselees' problems.

Making the Genetic Diagnosis

The interview. The first consideration of the genetic counselor is to find the answer to the question, "What brings you here?" Although people seek genetic counseling for many reasons, most often they boil down to a common problem: a certain disease has struck a member of our family. How likely is it to happen again — to us or to our children?

Having ascertained the reason(s) for the counseling, the counselor must next gather as complete a family history as possible; for ease of reference, he constructs a relationship diagram, known as a *pedigree*. This genealogical architecture is of the utmost importance. Several genetic diseases, most notably some of those affecting the nervous system, have more than one mode of inheritance. By revealing the numbers and relationships of affected individuals in a particular family, the pedigree often defines the correct inheritance mechanisms and therefore the recurrence risks.

In constructing a pedigree, the counselor notes parental ages: some chromosomal anomalies are more common with advanced maternal age, and some autosomal dominant mutations with advanced paternal age.

The genetic counselor is curious regarding the geographic backgrounds of his counselees. Different ethnic groups suffer from different genetic diseases, such as Tay-Sachs disease in Eastern European Jews, and thalassemia (Mediterranean anemia) in Greeks. In addition, careful inquiry sometimes reveals that "unrelated" couples in reality have ancestors from the same little village, thus suggesting that the couple may share common genes, productive of autosomal recessive or multifactorial diseases.

The causes of death and the current health of all relatives are of obvious importance. Of particular significance are childhood deaths. Many recessive inborn errors of metabolism have only recently been recognized as such, so that up to a short time ago, their victims were described as having died of pneumonia, convulsions, "failure to thrive," and so on. However, an accurate diagnosis can sometimes be made in retrospect at the counseling session.

The outcome of all pregnancies is thoroughly investigated. Multiple miscarriage is sometimes caused by chromosomal aberrations, especially translocations. In addition, single pregnancy mishaps may be significant. Dr. B. J. Poland and Dr. R. B. Lowry of Vancouver, B. C., recently demonstrated that about 10 percent of all examined stillbirths and miscarriage specimens had malformations which carried increased recurrence risks in subsequent pregnancies. These were most frequently chromosomal anomalies or severe maldevelopments of the nervous system. Certainly, a thorough autopsy should be performed on all stillbirths and, when possible, on miscarriages as well.

The physical examination. The *index case* or *propositus* is the family member in whom the genetic disease is first recognized. He should be thoroughly examined by the counselor and if necessary by selected adjunctive specialists. Proper recurrence risks cannot be given if the condition has been misdiagnosed. This is exemplified by the case of a married couple whose first-born child suffered from short-limbed dwarfism. A diagnosis was made of achondroplasia, an autosomal dominant condition. Since there was no other affected family member, the couple was told that their child's disease represented a new mutation; therefore, the recurrence risk would be no higher than that for the general population, a negligibly low figure. However, the second child was born with the identical disease. The astonished doctors told the couple that lightning could not strike a third time. But it did. At that point, the family was referred to Johns Hopkins Medical School, where Dr. Victor McKusick made the correct diagnosis of diastrophic dwarfism, an autosomal recessive condition with a 25 percent rate of recurrence in sibs.

It is often a good idea to examine all available family members, in addition to the propositus. Some genetic diseases are variable in their severity, for example the autosomal dominant Marfan's syndrome. Badly affected persons are quite tall, and they have thin fingers, bone anomalies, serious cardiovascular disease, and "slipped lens" of the eye. If a child is the only person in his family to be so affected, his case might be assumed to have resulted from a new mutation; thus, the risk to his future sibs would be negligible. However, if careful examination should reveal a "slipped lens" or a minor bone abnormality in the seemingly normal father or mother, then that parent must also carry the gene, and the risk for further children becomes 50 percent.

The laboratory work. Ancillary testing is sometimes of help in nailing down a particular diagnosis. The most frequently employed of these procedures are chromosomal analyses, biochemical studies, and X-rays.

Chromosomal analyses are usually performed on white blood cells. They are of the greatest utility when an infant or a child shows the clinical picture of a malformation syndrome. When a particular syndrome is due to a numerical or a structural chromosomal abnormality, knowledge of the precise chromosomal defect in the patient and study of the chromosomes of the relatives are of critical importance in defining the recurrence risks. For example, all patients with Down's syndrome should undergo karyotyping, since a small proportion of cases of mongolism are caused not by the usual trisomy-21, but by a familial translocation (see Introduction), which carries a much higher chance of recurrence. In such a situation, a woman was delivered of five consecutive children with Down's syndrome before the nature of the problem was recognized.

However, it should be recalled that chromosomal studies are of no use in the diagnosis of single-gene or multifactorial diseases. This fact frequently must be explained to patients and to their referring physicians, who may harbor exaggerated ideas regarding the usefulness of karyotypes.

Biochemical analyses for specific compounds are sometimes performed on blood serum, urine, or cultured skin cells. The presence of abnormal body chemicals or the absence of normal ones is characteristic of most of the inborn errors of metabolism. A biochemical abnormality of intermediate degree is diagnostic of the heterozygous carrier states for some of these diseases (see chapter 3).

X-rays are helpful in differentially diagnosing malformation syndromes, since many of them include more or less characteristic bony abnormalities.

Explaining the Problem and Making the Decisions

When the genetic counselor is satisfied that he has arrived at the proper diagnosis, he is ready to discuss both the risks of recurrence and the prognosis of the disease, and to help the family members arrive at their decisions.

Explanation of recurrence risks. The counselor first explains the manner in which the disease is inherited: chromosomal, autosomal dominant or recessive, X-linked, multifactorial, nongenetic, or unknown. Frequently, to clarify the different types of inheritance for the patient he uses diagrams like Figures 14-17 in the Introduction; then, reference to the pedigree reveals who is and who is not at risk to develop or to pass on the disease.

Quoted risk figures may be either specific or empiric. With single gene defects, *specific risks* may be precisely calculated from knowledge of the inheritance pattern: the chance that a child of a patient with an autosomal dominant disease will be similarly affected is exactly one in two, or 50 percent. However, the chance of recurrence of a chromosomal or a multifactorial disease cannot be estimated in this manner; therefore, statistical studies are made of large numbers of people at risk. In this way, it can be learned, for example, that approximately one of every 100 babies whose older sibs have Down's syndrome will suffer from the same disease. This is an *empiric risk* figure.

Risks should always be stated in absolute, rather than in relative, terms. It would be terrifying to be told after the birth of a child with cleft lip and palate (multifactorial) that the risk for subsequent children is increased forty times over the general population risk. But the perspective changes when it is realized that the general population risk is one in 1,000, while the recurrence risk after one affected child is one in twenty-five (forty per 1,000).

Another counseling technique to be recommended is the quoting of risks in obverse-reverse figures. Thus, the above couple might be told that the chance of their second child's having cleft lip and palate is 4 percent, but that the chance of his *not* having it is 96 percent. This may smack of lily-gilding, but persons at risk tend to develop tunnel vision, and such procedures help them attain more accurate emotional perception of the situation.

Three difficulties may prevent the accurate assignation of a recurrence risk. First, it is sometimes impossible to make a specific diagnosis of the disease, especially if the propositus is deceased. Second, a diagnosis may be established, but the condition may be so rare that the mode of inheritance has not been definitely established. And third, some genetic diseases are inherited by different mechanisms in different families, and the pedigree may not reveal which mode is operating in a particular case. In these circumstances, gross general estimate levels must be used.

Discussion of disease prognosis. Most recessive diseases are rather uniform in severity from patient to patient, while autosomal dominant conditions vary tremendously in expression. Chromosomal and multifactorial diseases are usually intermediate in this respect. These rules of thumb are taken into consideration by the counselor as he explains the likely course of progression of the disease in question, both in the propositus and in other affected and potentially affected persons. In the relatively few cases where treatment is available, the counselor describes it; where no therapy exists, he may suggest methods to minimize discomfort and disability.

The concept of *burden* is an important one. It is burden, rather than severity of disease per se, which appears to be the paramount factor in the decisions of couples regarding further reproduction. Burden is well illustrated by a consideration of the multifactorial traits anencephaly and spina bifida. (See chapter 5). Anencephaly is a developmental defect where most of the brain tissue is missing. No more severe disease could be imagined, but since the patients die at birth or shortly after, the parents rarely have any hesitancy about trying another pregnancy. However, spina bifida (failure of the spinal cord and vertebral column to form properly) is another story. Many of these patients survive for years, and an appreciable number have hydrocephaly ("water on the brain"), mental retardation, leg paralysis, and urinary system difficulties. Frequent, expensive surgery is the rule. Thus, the financial and emotional burdens of spina bifida are severe, both in terms of intensity and duration, and so the subsequent fertility of parents with such children traditionally has been drastically reduced. Fortunately, this is no longer necessary, since anencephaly and spina bifida can now be diagnosed prenatally, as will be described in chapter 5.

Helping the family to make decisions. This, of course, is the crucial step, the ultimate reason for all the preceding diagnosis and discussion. Currently, there is considerable argument concerning the proper role of the genetic counselor in this situation; this controversy will be discussed in the next section of this chapter.

Controversies in Genetic Counseling

A great number of the difficulties encountered by genetic counselors are related to the fear and loathing with which most people view genetic diseases. The medical term for a malformed newborn is "monster" or "monstrosity," and it is well established that such mythical monsters as centaurs, sirens, and griffins had their origins in descriptions of developmental malformation syndromes that are still being seen today. As in the past, the birth of such a child is often ascribed to divine retribution, so that when these unfortunates survive the newborn period, they are in many instances consigned to a life of misery and neglect in an attic or an asylum, lest the horror-struck neighbors socially isolate the family whose "bad blood" has come to light.

We fear the unknown. Everyone now takes for granted the fact that early diagnosis of cancer is desirable; yet sixty years ago, the distinguished Baltimore gynecologist Thomas Cullen received a great deal of abuse from his colleagues and from the laity, because he dared to publicly advance this belief. In 1914, cancer was never discussed openly, and patients with its symptoms stayed at home until pain and disability forced them to the hospital to die, while their relatives hastily concocted suitable lies to tell the neighbors.

Today, genetic diseases occupy a similar position, and so, much of the geneticist's time is spent in educating his counselees to look upon genetic disease in a more rational manner. It must be realized that genes are intrinsically neither "bad" nor "good." For example, consider the gene whose double dose causes the autosomal recessive sickle cell anemia. When present in a single dose, this gene actually decreases its carrier's likelihood of dying from malaria, a most useful trait for someone living in Africa. Our genetic diversity is necessary to evolution, and anyone's particular genetic constitution is a reasonable cause for neither shame nor self-exaltation.

The genetic counselor often must stress the inappropriateness of the guilt felt by so many parents after the birth of a deformed child. In addition, he has to be on guard against the common attempts to "put the blame" on the other side of the family. Once again, genetic disease is no one's "fault."

The Role of the Counselor in Decision Making

Virtually all genetic counselors agree that counseling should be directed toward meeting the needs of the individual, rather than those of society in general. Society might best be served, for example, were first cousins never to marry, since they possess a large number of genes in common, thereby increasing by a couple of percentage points the chance that they might have a baby with an autosomal recessive or a multifactorial disease. But this small added risk is usually not terribly frightening to an individual pair of affianced cousins, and it would hardly be proper for a counselor to try to talk them out of reproducing because "if everyone were to marry his cousin, there would be many more children with genetic diseases."

Genetic counselors are divided as to the attitude they should adopt regarding the decision-making process. Some counselors try to be totally nondirective: they simply present their "clients" with the objective facts, and will not offer anything that might be interpreted as advice. At the other end of this philosophical spectrum are the directive counselors, who also give their "patients" the statistical and medical facts, but in addition try to provide full interpretation. Some go as far as to suggest that it would or would not seem reasonable for their "patients" to run specific risks.

A third group of counselors adopt an intermediate approach. They feel that their special knowledge of genetic disease obligates them to go beyond mere risk-giving; thus, they try to describe to their "consultands" the impact of a particular disease on a patient and on his family. Some of these counselors may give an answer to the frequently asked, "What would you do in my place, Doctor?" but they're careful to point out the great subjectivity inherent in decision making, causing the doctor's answer to be necessarily satisfactory only for the doctor.

The third approach to counseling seems to me the most appropriate, but it must be admitted that no studies exist that could aid one in deciding this question at anything other than the gut level. We may eventually be helped by the research now being done by some teams of psychologists, who are subjecting the process of decision making to critical analysis.

Crucial to the decision-making process is the issue of withholding information. In line with the reasonable medical dictum "first of all, do no harm," many geneticists do not divulge information they fear will adversely affect the patient. As mentioned in the Introduction, this may be done when a newborn infant or a child is discovered to have XYY sex chromosomes. Similarly, some counselors do not tell the mother of a child with Down's syndrome about the association of aneuploidy with advanced maternal age. Probably the most universally observed genetic "conspiracy of silence" involves the so-called "XY female." These are relatively rare individuals with male sex chromosomes who have inherited either nonfunctional testis-producing genes or an inability to respond to male sex hormones. Therefore, they develop anatomically and psychologically as females. Although they are infertile and usually do not menstruate, their psychosexual orientation is entirely female, and in every sense of the word they are women. The fact that they have Y chromosomes does not make them men any more than having XXX sex chromosomes would make them "women-and-a-half." Yet, this information is almost universally withheld, for fear of causing psychological decompensation in the patients.

To my mind, this is especially pernicious reasoning. By such actions, geneticists can only reinforce the popular view of genetic disease as something more than a little unnatural, and better not mentioned. Far more appropriate would be a thorough explanation to the effect that all embryos start out with the capacity to develop into an individual of either sex, that many factors combine to direct this development, and that the makeup of a person's sex chromosomes is relatively unimportant: the truly significant matters are the adopted gender role and the anatomic sex.

An especially unsavory type of withholding of information is that done by obstetricians who refuse to tell their pregnant patients at risk for some genetic diseases about the existence of the new procedures for the prenatal diagnosis of these problems (see chapter 5.) The obstetricians behave in this fashion because they themselves believe prenatal diagnosis to be dangerous, immoral, or not thoroughly enough tested. There can be no defense to such a policy, whereby the patient is deprived of the option to become fully informed and then to decide for herself the course she wishes to follow.

The "Invasion of Privacy" Issue

One aspect of the "invasion of privacy" controversy involves the question as to whether relatives of counselees should be contacted, for the purpose of informing them of their own risk status. Since these are persons who did not seek counseling, the concern has been expressed that they may direct their dismay and resentment against the bearer of the bad tidings. However, Dr. M. A. Lubs of Colorado has now reported a study in which families related to patients with either hemophilia or muscular dystrophy were sought out and offered counseling. Well over 90 percent of the contacts were grateful that the effort had been made, and that they were given the opportunity to know whether or not they were at risk, so that they could employ their risk figures in their childbearing plans.

There is general agreement that it is reasonable to try to contact individuals at risk when the disease in question is preventable by prenatal diagnosis (for example, chromosomal translocations and X-linked diseases), or when the disease is treatable. An example of a treatable disorder would be Wilson's disease, an autosomal recessive condition characterized by the inability to metabolize copper, leading to cirrhosis of the liver, severe neurologic abnormalities, and death. The administration of a drug named penicillamine, however, causes the copper to be excreted from the body, thereby preventing damage and allowing the patient to lead a normal life.

On the other hand, there is tremendous disagreement as to whether patients at risk for incurable, nonpreventable diseases should be searched for. Autosomal dominant Huntington's chorea is such a disease. Certainly, it would be dreadful to discover that you stand a high risk of developing a fatal degenerative disease in your thirties or forties; it would in addition be frustrating to know that there exists no means to tell in advance of symptoms whether you will be affected, and no way to definitely avoid having affected children, short of not reproducing at all. Again, one can argue that the individual has the "right to know," but the problem is that he has not asked to know. Does he also have the "right *not* to know"? (This issue of "the right not to know" will receive additional attention in chapter 3.)

Sometimes, a counselee will insist that his relatives at risk not be contacted. Such a situation creates a serious dilemma. As the advocate of the person who has sought his help, the doctor is bound by tradition to respect the wishes of his patient. But does the patient's insistence on total confidentiality compel the doctor to leave those at risk in ignorance, especially when the disease in question is preventable or treatable? Whose rights and whose needs are paramount? This question is nowhere near resolution.

One possible solution might lie in the establishment of linked regional genetic "data banks," which would be used, among other things, for confidential pedigree storage and dissemination. Such a suggestion invariably brings forth loud cries of fear over the possibility of misuse by insurance companies, blackmailers, and Big Brother. However, tumor registries are standard practice both here and in Europe; in fact, the Norwegian Gynecological Tumor Registry has helped Dr. Per Kolstad and his staff to achieve the world's best results in the diagnosis, treatment, and research of these cancers. Abuse of information has not been a problem. I suspect, however, that had someone tried to set up a tumor registry sixty years ago, we'd have heard all the same arguments that are now being directed against genetics registries. The real issue is the fear of being exposed as having or carrying a "dread disease." Once we come to look upon genetic diseases as being no different from other categories of illnesses, I think we'll not mind being listed in a registry, if that will enable us to deal more effectively with our genetical problems.

The Abortion/Contraception Issue

The position of ethicists such as Paul Ramsey and Leon Kass has been stated: fetuses are human persons, abortion therefore constitutes murder, and genetic counseling directed toward this end is indefensible. The other side of the question is stated by Joseph Fletcher, who feels that to fail to prevent the birth of a deformed child is immoral and irresponsible. This controversy could be argued ad infinitum, but I won't do that; I couldn't change anyone's mind, anyway. Nor do I desire to. I merely hope we will all remain free to act upon the dictates of our own consciences with regard to this unresolvable, wholly subjective philosophical issue.

Techniques and Intellectual Capacities

Should the counselor modify his techniques according to the intellectual capacities of his counselees? Although it is currently fashionable to deplore medical paternalism, it seems falsely egalitarian and even a bit unfeeling to refuse to recognize that some counselees are hopelessly unable to comprehend principles of human biology or probability. No amount of study or patient instruction could help me to understand the workings of a giant computer, and I would be put out no end by a mathematician's detailed explanations when the limits of my capability extend only to being able to remember which switch turns the machine on, and where I should place the cards. Similarly, when dealing with an illiterate or a retarded counselee, it's usually futile to try to say more than, "Your child has what is called 'Wilson's disease,' which can be a very serious thing. But as long as he takes his medication, he'll be all right. Also, any other children you have are likely to get the disease, too; they'll all have to be checked out, so that you can see they get the best possible care." Then, you try to elicit questions, and conclude by arranging close medical and (if desired) contraceptive follow-up for the family.

Maximizing Benefits of Genetic Counseling

To date, there have been a few major studies attempting to evaluate the usefulness and efficacy of genetic counseling. Carter's group from London showed that most counselees understood well, and usually acted in a manner appropriate to the advice and to their own expressed desires. However, Leonard's Johns Hopkins team found that only half of their counselees had come away with a good grasp of their problems. Another quarter showed fair understanding, and the rest, poor. In fact, 8 percent of the patients denied ever having been counseled.

Thus, there seems to be room for improvement. Genetic counselors are always looking for methods by which they may impart their advice in a clearer fashion. In the long run, the increased teaching of genetics in high schools and colleges will help. But in addition, there is much that the counselee can do to obtain the optimal return on his genetic counseling investment.

The first prerequisite is to be certain that the counselor is competent. There exist no practice regulations, so that anyone who wants to call himself a "genetic counselor" is perfectly free to do so. There are three good reference sources to use in finding a qualified counselor. The first is your family physician, pediatrician, obstetrician, or other doctor. The second is the information service run for doctors and laymen by the National Genetics Foundation, 250 West 57 Street, New York, New York 10019. The third source is the National Foundation/March of Dimes *International Directory of Genetic Services*, edited by H. T. Lynch (fourth edition, March, 1974).

Once an appointment is made, the next step is to become familiar with pertinent genetic principles such as those described in the introductory chapter. Some of the references listed at the end of the Introduction may be of further help; in addition, the counselor may be able to recommend basic material pertinent to a particular problem.

The *bête noire* of genetic counselees is probability. This was evident in both Carter's and Leonard's findings and in the report of Dr. R. F. Murray's group at the 1974 meeting of the American Society of Human Genetics. As is the case with genetic principles, education seems to be the key to the probability problem. While I deeply sympathize with anyone who doesn't get along well with numbers, it nonetheless seems apparent that this is a situation that counselees should face squarely. To this end, I have included a couple of references on probability in the list for this chapter. King and Read's *Pathways to Probability* deserves special mention; the information in the majority of its 135 painless and charmingly written pages is intelligible even to me. If in the end a counselee can keep in mind only one probability fact, it should be: "Chance has no memory." If the risk of a disease in a couple's children is one in two, that risk holds *for each pregnancy*. After nine heads in a row, the chance that an honest coin will come up heads on the tenth throw is still one in two: no less. I'm sorry if that doesn't seem quite logical, but that's the way it is.

It's probably a good idea not to rush into counseling immediately after the birth — or death — of an abnormal baby. Leonard's information showed that the attendant emotional strain had a definite adverse effect on comprehension and later recall. Better to first give the mind time to work through the natural shock and grief. If, however, someone in such an unhappy situation feels he must have the information in order to achieve peace of mind, it would be wise to request a follow-up visit three to six months later. This arrangement serves both to check out the accuracy of recall, and to rethink the decisions that are so difficult to make in the heat of a tragic situation.

The counselor can be only as effective as he is permitted to be. A consultand should give him all the help possible. Material withheld because of embarrassment, shame, or any other reason will not be of use in clarifying a situation, and in fact, withholding information may create a pedigree with a "false inheritance pattern." The consultand should bring to the clinic all pertinent medical records, all available autopsy reports, and especially old family photographs. In our clinic, this has solved several dilemmas. One of our counselee-couples brought in their young child who was deaf due to Waardenburg's syndrome, an autosomal dominant condition. Besides deafness, patients with this disease may have white patches of hair and depigmented regions of skin. Examination of the parental couple was entirely normal, with the exception of a small light area of skin on the mother's back; this was not considered sufficient evidence for us to decide that she carried the Waardenburg gene with only minimal expression. However, she recalled having heard that her long-dead grandfather had always been "hard of hearing," but was touchy, and would never admit the fact. A hastily unearthed sepiatone print showed the old man standing majestically at the center of a family reunion, sporting a tremendous white forelock. Thus, the couple could be told that their son's disability certainly did not represent a new mutation, and that the risk of Waardenburg's syndrome in their subsequent children would be 50 percent.

Many families prefer to be counseled as a group, when the information to be exchanged is not of an unduly personal nature. However, anyone who would be made uncomfortable by this arrangement, for whatever reason, should have no hesitancy about requesting a private session. Comfort of the counselee should always take first consideration.

At the end of the counseling session, many geneticists try to ascertain whether the counselees have understood the situation by asking them to repeat the information. It's surprising how often this practice reveals a basic error in comprehension, which usually is easily corrected. Furthermore, having to explain data to someone else cannot help but fix the information more firmly in the mind of the explainer. When a counselor does not follow this practice, the counselee should ask permission to try to explain back the pertinent data. There's no reason for embarrassment. Remember, a garbled message can as easily be due to faulty transmission as to difficulties with reception.

In addition, counselees probably do well to request that a written report of the information be given them. This is not the usual procedure at most centers, but it may soon become routine. Dr. Y. E. Hsia of Yale found that this practice was greatly appreciated by patients, who frequently reread the information at home. Furthermore, the accuracy of recall of these patients was rated as 75 percent good and 25 percent fair, a most favorable record. It's easy to see how possession of a written record would prevent confusion such as that which led one of Fraser's patients to convert a one-in-four risk for an abnormal child to a one-in-four chance of a normal one.

One of the real problems of genetic counseling is that the information given often creates considerable worry over the possibility of having an abnormal child. Leonard's data revealed frequent severe maladjustments of sexual relationships when proper contraceptive counseling was not provided. The solution to this problem is obvious: genetic counselors must be careful to ascertain that their counselees have access to competent and sympathetic gynecologists, urologists, or family practitioners. And any counselee in need of such a referral should ask for it.

Finally, counselees ought not to fret over the difficult decisions that so often accompany genetic counseling. Geneticists haven't invented these "new" diseases; they've simply recognized them and in many cases made it possible to prevent or even treat them. The capacity to make reasonable decisions is the essence of being human, and is preferable to the ostrich technique, which, while protective of the head, still leaves a considerable portion of the anatomy exposed to painful attack.

We have seen in this chapter how diagnosis and explanation furnish the education necessary to permit counselees to take a rational role in the individual management of their genetic problems. In the remainder of the book, I'll discuss the current and future preventive and therapeutic options that may follow upon genetic counseling.

2

Eugenics, Euthenics, and Euphenics

Eugenics

Eugenics is a dirty word. For most people, it conjures up images of goosesteppers and gas chambers. This is curious, for the word takes its origin from the Greek *eugenes*, meaning "well-born." How, then, did it come to acquire such unpleasant connotations?

Ultimately, I think that the paradox is based upon the fact that eugenics may be defined as the attempted application of genetical knowledge to the improvement of the human race. When we forget about the individual in the interest of "the race," we usually get into trouble. So it has been with eugenics.

Historical Development and Definitions

The word "eugenics" was coined in 1883 by Francis Galton, the great British mathematician and statistician. Galton was a compulsive measurer and, in looking for new worlds to conquer, he eventually turned his attention to the attempted quantitation of human health and personality traits. The Austrian monk Gregor Mendel had published his epochal studies on the mechanisms of gene inheritance in 1865, but the work remained in obscurity until 1900. Thus, having no knowledge of basic genetic transmission data, Galton formulated his concepts involving the "blending" of biological traits in the offspring, such that they could be analyzed mathematically. This was the essence of the science of biometrics.

In those days, it was vaguely assumed that the hereditary factors were somehow transmitted via the blood; thus, the terms "blue blood" and "bad blood." Despite this erroneous thinking, humans had for millenia been selectively breeding their plants and domestic animals for desired characteristics, and Galton made a careful study of the techniques of selective breeding. He also gave some thought to human breeding, and he noted that distinctive accomplishments seemed to run in families. This, in fact, was amply illustrated by his own pedigree, which included numerous men of great scientific distinction, including his cousin Charles Darwin.

In such fashion did Francis Galton finally reach the conclusion that encouraging superior persons to mate might improve the general character of the human race. This idea is called *positive eugenics.*

A novel thought? Hardly. Utopian societies were all the rage in the United States during the second half of the nineteenth century, and many of them tried to breed their children "scientifically." For example, the Oneida Community of New York, founded in 1851, determined their matings by committee decision; a feature worthy of comment was that a high number of offspring were sired by John Humphrey Noyes, the founder and "spiritual leader" of the community. Due to racking internal dissension, some of the causes of which should not be beyond our imaginations, the Oneida Community had to disband in 1881. So much for Utopia in New York.

Galton's call for intensified propagation by the better classes won him some followers, and resulted in his appointment as Director of the Laboratory for National Eugenics at the University of London. Nevertheless, his tenets did not meet with overwhelming general approbation. Shortly before his death in 1911, he formed the opinion that the genetic millenium might be more rapidly attained were he to incorporate an aspect of *negative eugenics,* discouraging the reproduction of poorer specimens of mankind. Thus, by a judicious combination of positive and negative eugenic measures, Galton hoped to enhance the effects and mitigate the harshness of the process of natural selection.

It frequently happens that the followers of great innovators are themselves possessed of much narrower minds. Such a man was Galton's successor at the Eugenics Laboratory, Karl Pearson, whose teachings directly foreshadowed the dreadful misapplications of negative eugenic principles in Western societies during the first half of this century. Pearson devoted a good deal of worry to an imagined populational decline in physical and intellectual characteristics; the only answer, he thought, lay in discouraging reproduction among those with hereditary defects. He doubly compounded the felony by pooh-poohing Mendel's laws of genetic inheritance, and denying the importance of environmental factors in the development of the individual. Thus, he listed as genetic diseases such items as tuberculosis, pauperism, and criminality. Pearson's writings and speeches clearly showed the disgust and contempt he felt for "genetic undesirables."

The German philosopher Friedrich Nietzsche shared Pearson's attitudes. In *Thus Spake Zarathustra,* Nietzsche called for the breeding of a race of supermen. Note how one bit of sloppy thinking leads to another, and yet another: from breeding plants and animals with a specific desired trait, to breeding humans with increasingly superior general genetic endowments, to the elimination of disease by properly directed reproduction, to breeding a superrace.

It was not long before these fine European exports had reached our shores. By the first decade of the twentieth century, they were fanning the fires of Social Darwinism, a concept that had already enjoyed considerable popularity in America for two decades. Social Darwinism was founded on the naturalistic philosophy of Herbert Spencer, who proposed analogies between societies and biological organisms with regard to Darwin's concept of "the survival of the fittest." Thus, material success in business or a profession was seen as the natural and logical consequence of (hereditary) biological superiority. An important precept of Social Darwinism was that the unsuccessful, by whatever criteria, are intrinsically (genetically) inferior, and so deserve to die out. The corollary is that the successful must by definition be intrinsically superior, so that in the context of societal evolution, wisely directed breeding should lead to ever-superior creatures. The novelist and story writer Jack London was an enthusiastic Social Darwinist and Nietzschean; this philosophy strongly influenced his later works. In addition, journalist Henry Mencken professed to admire the doctrines of Social Darwinism, but his sense of humor and his basic belief in individual rights sometimes led him to pen contradictions of the sort one might expect from someone who hates humanity but loves people.

Enthusiasm for Social Darwinism combined with the post-war mood of disillusionment, isolationism, and xenophobia. The result was an especially nasty chapter in the history of American eugenics. American eugenicists had early broken away from Galton's biometrics, going to the opposite extreme of overenthusiastically embracing Mendel's Laws of single-gene inheritance. They were convinced that all human traits were determined by the action of single gene pairs; for example, the eugenicist Charles B. Davenport never abandoned the opinion that all forms of mental retardation represented an autosomal recessive trait and that the tendency to become angry was an autosomal dominant. Furthermore, these eugenicists believed that the genetic composition of different races was vastly different, with the Nordic branch of the Caucasian race having the fewest physically and socially undesirable genes.

By 1920, the leaders of the American eugenics movement were the above-mentioned Dr. Davenport and Dr. Harry Laughlin, both of the Eugenics Record Office at Cold Spring Harbor, New York. This organization had the zealous support of a large group of hopeful reformers, misguided scientists, and blatant racial bigots such as Madison Grant (author of "The Passing of the Great Race"). Laughlin and Grant led the drive to convince our legislators that unrestricted immigration by undesirable ethnic groups (non-Nordics) would certainly lead to pollution of American society, with an increase in crime and immorality. They found the congressmen remarkably willing to swallow their pseudoscientific data "proving" that immigrants from regions other than northwestern Europe were bringing into America large numbers of deleterious genetic traits. An alarmed Congress moved appropriately, and the result was the United States Immigration Act of 1924, one of the few items which produced an actual display of enthusiasm from President Calvin Coolidge. Until the Celler Act of 1965, our immigration regulations were strongly influenced by Laughlin's and Grant's unscientific and vicious racist propaganda.

A second shameful feature of American eugenics was the passage of compulsory sterilization laws, another result of the endeavors of the energetic Dr. Laughlin. Although court-ordered sterilization of retarded persons was not a new idea (as witness the famous *Buck* vs. *Bell* case of Justice Holmes), Harry Laughlin sought to move the emphasis from the inefficient consideration of individual cases to the more convenient generality of entire subpopulations. Furthermore, Laughlin's targets included not only the mentally retarded, but also a great range of "socially inadequate" persons, including criminals, epileptics, drunkards and drug addicts, those suffering from chronic infectious diseases, the blind, the deaf, the deformed, and the insane. Laughlin even included a category called "dependent," consisting of ne'er-do-wells, tramps, paupers, and orphans. Although more than half of our states passed genetic sterilization laws, these statutes have never been vigorously enforced, and a United States District Court decision has recently struck them down as unconstitutional. That this may represent a mixed blessing for retarded individuals, I will consider later in this chapter.

Fortunately, the American proponents of negative eugenics never went so far as to advise killing the "socially inferior," nor did the proponents of positive eugenics advocate the governmental control of marriage and mating. But, as we all know, both these practices were part of the official policy of the pre-war and wartime Nazi government of Germany. It should be a cause for sober reflection that such practices could occur with at least the tacit approval of the overwhelming majority of the citizens of a nation which until shortly before had led the world in artistic and scientific achievements. Especially should it cause us to look with skepticism on the pronouncements of those who take early, fragmentary scientific data, use them as a basis to predict future developments in genetics, and then offer us strongly worded advice as to how we may evade our otherwise inevitable genetic difficulties.

By the late 1930s it appeared that the American eugenics movement had pretty well run its course. This was due partly to popular revulsion with the Nazi perversions of genetics, and partly to other factors. For one thing, the economic disasters of the thirties made it apparent to even the most rabid boosters of Social Darwinism that material success could not be considered a hereditary trait. In addition, genetic scientists had accumulated and publicized data that made basic eugenic proposals appear ridiculous. For example, the mathematical demonstration by the British geneticist Lionel Penrose that mental retardation could not be determined by the effect of a single gene deprived the movement of one of its most cherished goals.

However, by the late forties, eugenics began to show some phoenixlike behavior. There were some important basic changes: the approach was more compassionate and humanitarian, and appeal was made to reason and voluntarism rather than to force and legislation.

One of the early leaders of the modernized eugenics movement was the highly regarded geneticist H. J. Muller. Dr. Muller had received a Nobel Prize for his investigations of gene mutations in the fruit fly Drosophila. That work together with his continued interest in and study of mutations led him to the conclusion that we will sooner or later come to carry an intolerable load of mutations, and thus will face a "genetic twilight." Muller based his gloomy prognostications on our increasing exposure to mutation-causing radiation and chemicals, and our progress in medical knowledge and technology which was allowing us to save the lives of many sufferers of previously fatal genetic diseases. As the primary solution to the dilemma, Muller proposed voluntarily increased reproduction by those in good genetic health; as the major means to this end, he advocated the widespread adoption of *eutelegenesis,* a concept delineated by Herbert Brewer in 1935. Eutelegenesis involves the insemination of women with sperm from "superior" men, and Muller envisioned regulatory boards that would store sperm from our great and near-great, making it available to women wishing to become pregnant in such a fashion. This interesting concept will be further discussed in chapter 7.

Most modern eugenicists share Professor Muller's concern for the health of our collective "gene pool"; they call on us to be less selfish and to consider the needs of our descendants. Dr. James F. Crow of the University of Wisconsin states, "I am an ardent conservationist . . . I am quite willing to put severe restrictions on this generation in order to have a better environment next generation. I am quite willing to put restrictions on individual freedoms this generation in order to have a lower mutation rate for the benefit of our posterity."[1]

In the same vein, Dr. Cedric Carter of England sees family planning as a hopeful basis for future positive eugenic action. He points to the fact that fertility rates in the upper social classes seem to be showing a relative increase, and he urges such persons "not to be too humble about their own gifts and those of their children, and to have the courage to be nonconformist when the two-child family was fashionable."[2] In fact, he suggests that "the more energetic and gifted parents" should have at least four children. (Aside from the definition of "gifted," one might make argument over how long such a couple could possibly remain energetic.)

Gottesman and Erlenmeyer-Kimling deplore the traditional emphasis on IQ enhancement as the aim of eugenics. Instead, they suggest the use of an Index of Social Value, for which they provide an abbreviation (ISV), but no definition. They state that the truly important factor is the "social value" of different individuals, and they would leave the definitions and decisions related to this term to a committee which would be heterogeneous with regard to social class, race, and intelligence.

Some of the present-day eugenicists are eager to put the techniques of reproductive engineering to eugenic use. Muller's advocacy of eutelegenesis is an example of this; in addition, cloning and *in vitro* fertilization have been held out as eugenic maneuvers of considerable potential. Even organized programs of prenatal diagnosis of genetic disease have been recommended as a possible means of improving the race.

Thus, although the modern and futuristic eugenicists all deplore Nazilike negative eugenics programs, the fact remains that their interest still lies in the improvement of mankind at large, rather than with the needs of individuals. Their redeeming feature is that they wish to rely upon persuasion rather than force as the means to the end. This is highly laudable, but let's not forget that the original Galtonian eugenics was also conceived as a purely voluntary program.

Controversial Aspects of Eugenics

The individual vs. *the race.* This central issue is much apparent in all the foregoing material. Ultimately, it relates to the fact that by definition, eugenics is concerned with the human race at large, to whose interests those of the individual must if necessary submit. We have to decide whether we are to act as genetic ecologists as Dr. Crow would wish, or whether we should take a more immediate view. I do believe that few of us would willfully worsen the genetic prospects of our remote descendants; however, even among eugenicists, there exists a marked diversity of opinion as to the proper course to pursue to avoid a "genetic twilight." This is primarily a reflection of the fact that no one can accurately predict the effects that any genetic measure may have, many generations into the future. All such considerations are of necessity vague and theoretical, a situation which does little to persuade many thoughtful persons to suddenly and radically limit their reproductive options. The mass of humanity is a conservative organism, and although this can sometimes be highly exasperating, perhaps overall it's a good thing.

The issue of sterilization of the mentally retarded deserves special attention. Fifty years ago, it was believed that this would go far toward reducing the number of defectives in society. California led the way by sterilizing around 10,000 such individuals. However, there exists no evidence to prove or even suggest that this has raised the mean IQ of Golden Bears by even a single point. Such lack of correlation becomes understandable when we realize that new mutations and fresh unfortunate multifactorial combinations are continually producing considerable numbers of persons with reduced intellectual capacities. In addition, both Carter and Penrose have now demonstrated that people with the lowest IQ levels produce the fewest offspring — not the greatest, as was previously thought. Penrose, in fact, made a point of the fact that "propagation of the unfit" is intrinsically ridiculous, since by definition the unfit are those who are incapable of reproducing. (The corollary of this statement should give us additional food for thought.)

But is this to say that the mentally retarded must not be sterilized? I don't think so. The truly valid question would seem to be not whether the sterilization is in the interest of society, but whether it is in the interest of the retardate. It is clear to me that one does not need a high IQ to be a good parent and to enjoy a rewarding family life, but there do exist persons who are sufficiently retarded that they are not capable of providing anything resembling proper child care. In addition, effective long-term contraception is a major problem for such people. Thus, sterilization could free severe retardates to enjoy sexual activity without the risk of what for them would be highly stressful and unhappy consequences.

Unfortunately, however, sterilization (and abortion) of the retarded are currently attended by serious difficulties. The problems are well illustrated by a recent case from our Genetics Clinic. A retarded couple in their late thirties presented with the woman's legal guardian. The husband and wife were capable of carrying out simple tasks under close supervision; as such, they helped to give nursing care to bedridden old people. They demonstrated a warm, mutually supportive relationship, and were maintaining an active, if not hectic, sex life, which they freely admitted constituted their only meaningful form of recreation. Despite intrauterine contraception, the woman had conceived. The couple and the guardian agreed that abortion seemed the preferable plan, because of an increased risk of multifactorial retardation in the offspring and because the wife was not thought to be capable of rearing a child. However, our legal counsel advised us that we could not legally abort or sterilize the woman. This advice was based on a 1974 United States District Court decision to the effect that no one may be aborted or sterilized without first having given free, informed consent, and that since minors and incompetents are by definition unable to give such consent, they may never be sterilized or aborted. Furthermore, no one may give consent for them — not guardians, parents, or judges. Thus, we found ourselves in a Catch-22 situation. The couple went home unaborted and unsterilized, their marital relationship badly threatened. The federal judge had decided that everyone should have the right to reproduce, but the principals in our case were quite unhappy that they did not have the right *not* to reproduce. Thus does the pendulum of policy swing from extreme to extreme, from unsavory compulsory sterilization laws to irrational restrictive sterilization laws.

Also in relation to the question of the race vs. the individual, one may question the whole idea of proposing eugenic concepts in conjunction with genetic counseling. In the past, the primary function of genetic counseling was to advance eugenic programs, so that even today, many prospective counselees refuse to come to clinic because of the fear that "they only want to tell me — or force me — not to have children." However, the presently declared goal of genetic counseling is the improvement of the lot of the individual (or the couple), whereas the basic goal of eugenics is still the improvement of the human race. Therefore, any accord between the two can be only coincidental. Genetic counseling is often *dysgenic* (as opposed to eugenic); this is illustrated by the discussion of cousin marriage in the previous chapter.

Another circumstance where genetic counseling can have a dysgenic effect is in the prenatal diagnosis of genetic diseases. Although it may seem that the abortion of affected individuals is eugenic, this is not the case, because *reproductive compensation* usually occurs: the parents replace aborted affected fetuses (whose fertility usually would be greatly reduced) with healthy asymptomatic gene carriers. The end result, then, is a net increase of carriers in the population. Yet, few genetic counselors would advise against cousin marriage or prenatal diagnosis on the grounds of responsibility to future generations. We do, however, sometimes see counselees who themselves express the thought that perhaps they'd do well not to reproduce, so as to help "wipe out bad genes." They must then be reminded that bad genes may also be good genes, that everyone carries a few "bad" genes, and that what with the constant occurrence of new mutations, their personal sacrifice would be a meaningless gesture.

Difficulties in selection of traits for continuation or eradication. A major problem in this sphere is the subjectivity involved in evaluating traits. You need do no more than go to a P.T.A. meeting to discover that some people want their children to be highly competitive, while others place a major premium on noncompetitiveness. Some parents hope for children with great athletic skills, others desire that their offspring excel at school, and still others feel that their children will be best off as gentle, kind, and unselfish mediocrities. In our more rational moments, however, most of us will agree that it's really proper for the children themselves to decide which traits to emphasize and which to de-emphasize. Perhaps we could do worse than to extend the same courtesy to all future generations of individuals.

Virtually all eugenicists have advocated that desirable traits be selected by special committees. But just as no one can agree on the relative desirability of different traits, no one can agree either on how to organize the committees with proper respect for wisdom and fairness. I myself would not care to have my reproductive habits dictated by any imaginable committee.

Even if we could agree on which traits to breed in and which out, it seems extremely hazardous to try to project far into the future which genes will be beneficial and which harmful, under environmental conditions that may be very different. What's desirable today may be deadly tomorrow. Suppose that thousands of years from now, we should be visited with a true-life version of Jack London's "Scarlet Plague." If we had bred for noncompetitiveness, the few survivors of the epidemic would probably die of starvation or exposure in a gentle and peaceful fashion. On the other hand, if we had selected for competitiveness, the survivors, unable to agree among themselves and in a fierce power struggle, might kill each other off. Once again, perhaps we'd do best to try to preserve our genetic heterogeneity.

Another problem relates to the fact that most traits considered under eugenics programs are multifactorial (involving polygenic inheritance), not single gene characters. Therefore, breeding them in and out would be a difficult, tedious affair, because of the large (and unknown) numbers of gene pairs involved. We know from animal and plant husbandry that it is quite unusual to be able to select for a good trait without having some bad ones crop up as well. The problem is exemplified by the decreased fertility seen in specially bred hornless goats and broad-breasted turkeys. Selection for increased egg size in chickens has been associated with decreased numbers of eggs. Selection for high yield in both Irish potatoes and American corn has been associated with lowered resistance to fungus infections, with resultant severe crop damage.

The unexpected appearance of undesirable traits is due to two mechanisms. First, a single gene may have more than one effect on the organism: this is called *pleiotropy*. Second, genes for a harmful trait may be closely *linked* to the genes for a desired trait. This means the two sets of genes are located close together on the same chromosome, and therefore "travel together," making it impossible to select for one without the other. The genes for a desired trait may be linked to genes whose deleterious effect is exerted by a recessive or a multifactorial mechanism. As long as the deleterious genes remain in the heterozygous condition, they cause no trouble; however, increased inbreeding (consanguinity) provides these genes increased opportunity to pair up and become unmasked, thereby making the organism worse in some respects, while it is improving with regard to the desired trait.

At this point, it might be interesting to consider a concept that is now causing a great deal of sensational speculation. Some people suppose that our increasing genetic knowledge will soon permit us to breed superathletes. According to W. E. D. Stokes, "Why, there is no trouble to breed any kind of men you like, four feet men or seven feet men — or, for instance, all to weigh sixty or four hundred pounds, just as we breed horses. It only takes a longer time and more patience."[3] This was written in 1917. The fact that such a claim and many others like it remained utterly unfulfilled for more than fifty years did not stop columnist Nicholas von Hoffman from hysterically alleging similar potentialities for reproductive engineering techniques. In 1972, von Hoffman wrote, "It's three or four years down the road yet, but. . . . You own a pro football team and you need a Joe Namath? A little salt, a little pepper, and a little something else and you got [sic] a super-Joe Namath."[4] Von Hoffman worried that this sort of business will cause all NFL games to end in ties.

I'd certainly be willing to give von Hoffman his full four years, and bet him that by the end of them, we'll be nowhere near being able to mass-produce athletic heroes. And for several reasons.

For one thing, in breeding a superathlete, which quality do you select for? Size? Strength? Speed? Good reflexes? Competitiveness? Each of these traits probably has a polygenic basis, and therefore, due to the large numbers of genes involved, it would take generations on generations to breed in enough "good genes" to significantly improve your desired trait. And of course, the more multifactorial qualities you want, the more you magnify the problem. Then, too, there's the pleiotropy-linkage double whammy. How do you know that after you've gone to all the trouble of selecting for athletic ability, you haven't also inadvertently selected for severe multifactorial bookwormism, or even recessive blindness?

Von Hoffman states in his article that *in vitro* fertilization could be used to give us our superathletes. But, in fact, whether fertilization were to occur in a laboratory vessel or by natural techniques, the genetic recombinations would be the same. Von Hoffman also thinks that the Athlete of Tomorrow might be produced by mass cloning. However, the genetic makeup of a clonee would be identical to — no better than — that of his nucleus-donor "parent." Thus, there really seems to be no technological alternative to the droll prospect of expensive human stud farms maintained by present-day businessmen, so that their successors several hundred years in the future may have quarterbacks who can throw the ball a little farther. That is, if they don't happen to be blind. And if people then are still playing football.

Parenthetically, if von Hoffman has watched Joe Namath's performances on different Sundays, he must realize that even if there were two entire teams of Namaths opposing each other on a particular day, it's unlikely indeed that the game would end in a tie.

In summary, I think the blather over the designed production of any kind of genetic superstar represents the purest kind of pie-in-the-sky. Like most talents, athletic ability is multifactorial: that presupposes the importance of such environmental factors as good food, wise training regimens, and encouragement. We'll get better athletes (and I suspect other benefits as well) by trying to improve our environment, rather than by attempted genetic manipulations. And all the newspaper and magazine space given over to silly speculations about the production of superathletes and superintellects might be better used for pleas to develop techniques to ameliorate the effects of the different genes that cripple the minds and bodies of so many of us.

The reproduction race. A strong argument against the implementation of eugenics programs is the certainty that whoever may be the crucial committee members, they will not decide to breed out their own traits, but those of others. This thought could make people nervous — and it does. Traditionally, blacks have fared poorly under state sterilization laws, and this fact has led many black leaders to oppose sterilization for blacks under any circumstances. These people also see contraception as a "white man's plot"; they urge their followers to forego birth control and have as large families as possible. Such advice frightens many unthinking whites, who react by increasing their own birth rates "in self-defense." The upshot of all this is that these individuals, both black and white, who are overwhelmingly poverty stricken to start with, end up with hordes of children they can neither feed nor educate properly, and so the cycle repeats for another generation. This is one instance where the interests of all races — black, white, and human — as well as the interests of most individuals would best be served by the same process: a reduced birth rate.

Conclusions

As must be apparent from my foregoing statements, I think it is not reasonable to recommend the adoption of eugenics programs. While I certainly agree with Dr. Crow that it is ethical to be concerned for the future, I'd have to add that it seems more than a bit foolish and even potentially harmful to plan man's genetic future without really knowing how to select for multifactorial traits, how to prevent concurrent expression of deleterious traits, and which traits truly will be of use or of harm in the environment of thousands of years in the future. Furthermore, the dire predictions of "deteriorating gene pools" and "genetic twilights" should be looked on with a jaundiced eye. We must remember that sixty years ago, Karl Pearson was making similar claims about increasing numbers of defectives, and the results included immigration restrictions, sterilization laws, and gas chambers. Now, Penrose's data indicate that there has been no decline in the general intelligence level.

In addition, the calculations of geneticists, such as Drs. Arno Motulsky and George Fraser, show that improving the fitness of persons with genetic diseases would not cause a significant rise in the numbers of affected individuals in the general population. For example, if all patients with a lethal autosomal recessive condition were to be cured and then reproduce, there would be about a one percent increase in the number of patients over the next five generations. For an autosomal dominant disease, there would be a sixfold increase over five generations. But since all these conditions are rare, advances in medical treatment are not likely to cause a significant absolute increase in the number of affected persons such as would overwhelm medical facilities.

Almost everyone has some traits which he would do well to pass on to the next generation and some which he'd do better to keep to himself. There appear to be no compelling arguments for us to surrender any of our individual reproductive prerogatives; indeed, it would seem wiser to retain our full complement of genetic heterogeneity as a hedge against the uncertain future.

Euthenics

Euthenics may be defined as attempts to improve the condition of the human race by manipulating the environment. Although euthenics is often considered to be analogous to eugenics, it is really a different concept. In euthenics, the emphasis is placed on minimizing the presence and the effects of harmful or unfavorable environmental situations, and so, most euthenic measures consist of attempts to help individuals make the most of what they've got. Moreover, the few euthenic procedures specifically designed to "help the race" simply boil down to attempts to make not only the future but also the present environment safer to live in.

Some Examples of Euthenic Measures

Experiments performed on the fruit fly Drosophila (man's best genetic friend), largely by Dr. H. J. Muller, have shown clearly that radiation and certain chemicals are potent producers of gene mutations. Since Drosophila DNA and human DNA are identical, it's logical to assume that radiation and chemicals are also mutagenic in humans. And since the large majority of mutations are harmful, it then follows that it would seem wise to limit our exposure to mutagens. This logic has become part of current popular thought, being largely responsible for our infatuation with "natural" items. Sometimes, this is carried to the extreme, as illustrated by the occasional cancer patient who refuses lifesaving radiation therapy so as not to risk gene mutations. Such cases to the contrary, it seems prudent to attempt to prevent unnecessary contamination of our surroundings by radiation and mutagenic chemicals.

Most other euthenic endeavors are individual affairs. Depending upon the severity of the underlying genetic disorder, they are more or less successful, but, I think, invariably praiseworthy.

Prostheses may be employed. In prehistoric times, severe myopia was a fatal genetic disease: nearsighted hunters could not get sufficient food to survive, and nearsighted women would have had considerable trouble taking care of the children and the cave. Today, eyeglasses and contact lenses have reduced myopia to the level of a minor nuisance. The list of prostheses is exhaustive, and some of them ameliorate genetic conditions which if untreated would be extremely disadvantageous in our present-day society: for example, the ingenious artificial arms that permit victims of many inherited skeletal disorders to lead normal lives. Nonprosthetic mechanical aids are also frequently helpful. One of my present patients suffers from diastrophic dwarfism; her tiny, deformed legs have never permitted her to walk. But her motorized wheelchair has allowed her to obtain a master's degree in social work, and to hold a responsible job. She is now pregnant, and since diastrophic dwarfism is inherited by an autosomal recessive mechanism and her husband is of normal stature, she has every reason to expect a "normal" child.

Drug therapy is another euthenic tool. Diabetes is a polygenic disorder that was invariably fatal until the development of preparations of injectable insulin. Growth hormone is used to permit attainment of normal height by individuals who are genetically incapable of synthesizing this compound. Antibiotics prevent innumerable genetic deaths, from the cystic fibrosis victims who would otherwise die early from pneumonia, to those "normal people" who have increased genetic susceptibility to one or another microorganism.

Another laudable euthenic program is the individualization of educational goals. This involves getting away from the insistence that a boy or a girl with exceptional athletic ability but little scholastic aptitude or interest must nevertheless go to college. Educational individualization includes special programs of education, so that a child with a reading disability may get not only assistance with this problem, but also encouragement to develop skills in areas where he may excel, such as mathematics or art. Few of us are totally devoid of strong points — or weak ones, for that matter.

The conquest of Rh-disease of the fetus may be considered the masterwork of euthenics. Since it is also the major triumph to date in the field of genetic therapy, it will be discussed in chapter 4.

Dysgenic and Eugenic Aspects of Euthenics

As I've already mentioned, highly informed and intelligent people such as Dr. H. J. Muller have expressed considerable concern over the fact that modern medical advances have made it possible for the victims of many genetic diseases not only to live normal lives (which the eugenicists don't regret), but to reproduce as well (which they do regret). Thus, eugenicists object in a sense to euthenics because it is dysgenic. That this is a relative matter is illustrated by the fact that no eugenicist in his right mind would want to breed out the genes for myopia. But there do exist physicians who consider it wrong for diabetics to reproduce, since there is an increased likelihood that their children will be similarly affected.

However, this is not a cut-and-dried matter. Eradication of malaria by mosquito control may be considered euthenic. But it is eugenic, not dysgenic, since in the absence of malaria, the "sickle-cell gene" loses its relative advantage and so will tend to decrease in frequency. The (euthenic) therapy of Rh-disease of the fetus is sometimes eugenic and sometimes dysgenic. As will be explained, all affected fetuses are Rh-*positive;* the mothers are Rh-*negative,* and the fathers Rh-*positive* (like the fetuses). Thus, when an affected male fetus is saved, the therapy has been dysgenic, but when the fetus is female, it has been eugenic. Such situations drive contemplative geneticists and eugenicists to fall to the floor and pull their hair out by the roots.

Euphenics

Dr. Joshua Lederberg, Nobel Laureate in Genetics and Professor of Genetics at Stanford University, is a moderate, thoughtful, and prolific writer in the field of Genetic Engineering. He considers that our knowledge of inheritance mechanisms is too fragmentary to justify implementation of traditional eugenics projects, but that in addition, euthenics is too often a matter of too little too late. For these reasons, he proposes an alternative concept which he calls euphenics. In essence, Lederberg's euphenics would include all measures that can be taken to avert the consequences of genetic difficulties, with the understanding that better results are achieved the earlier in the development of the individual that ameliorative action is taken.

Euphenics is seen, then, to include euthenics, but also many measures which might be considered individually "eu-genic," that is, designed to improve the genetic constitution of a particular person, rather than that of the race. Euphenics would constitute total biological and technological engineering, undertaken with the aim of helping everyone to realize his full potential for health and accomplishment. Thus, it would be specifically oriented to the welfare of individuals, not to that of society. Measures such as genetic screening, genetic therapy, prenatal genetic diagnosis, and all the novel techniques of reproductive genetics have euphenic potentialities, as related to their capacity to improve the lot of the genetic counselee.

Conclusions

Advocacy of eugenics programs would seem unjustified, in view of the great difficulty in prognosticating which traits should and should not be perpetuated, the incomplete state of our knowledge regarding the inheritance of all personality traits and many physical characteristics, and the demonstrated potential for major abuses of the legislation necessary to make certain general acceptance of these schemes. The use of genetic counseling to advance the supposed interests of the human race would seem improper and perhaps even unethical.

At this time, euthenic manipulations appear to offer the best means to help individuals deal effectively with their genetic shortcomings.

However, as our expertise increases with regard to both the techniques of genetic prenatal diagnosis, screening, and therapy, and the modalities of Reproductive Engineering, we should be able to follow the diagnostic and educational functions of genetic counseling with individualized euphenic programs designed to improve the lot of genetic counselees by manipulating their hereditary material as well as their environments.

3

Genetic Screening

Genetic screening is the systematic examination of populations, the purpose of which is to detect persons who carry hereditary abnormalities capable of producing disease in those persons themselves or their descendants. This definition includes a wide assortment of procedures. Genetic screening may be performed in order to detect genetic diseases in their early stages, or to identify healthy carriers of potential disease-causing genes. Screening programs may involve the general population, or they may be selective: that is, directed at specific subpopulations known to be at high risk. Depending upon the genetic condition under study, populations of specific ages may be screened. Fetuses, newborn infants, children, or adults all may come under scrutiny under different circumstances.

Ideally, genetic screening should not be performed unless four conditions can be met. First, the test used should be technically simple, inexpensive, and suitable for mass processing. Second, there must be adequate provision for following up positive or suspicious results, both to verify them and to give the patients appropriate test interpretation and subsequent counseling. Third, test reliability must be extremely high. A certain number of false positive tests would be permissible, since follow-up testing would rectify the errors. However, false negative results, leading to nonrecognizance of affected persons or carriers, could not be tolerated. Fourth, there must exist a reasonable method of treatment or prevention for the disease under screening; otherwise, there is no practical purpose to the program.

In this chapter, I will consider genetic screening by dividing the subject into four parts: screening for single-gene defects (including autosomal recessive, autosomal dominant, and X-linked recessive diseases), screening for multifactorial defects, screening for chromosomal disorders, and controversial and futuristic considerations.

Many screening procedures for single gene disorders involve

testing for a specific gene product, such as an enzyme whose absence is responsible for the genetic disease in question. Alternatively, the genetic screener may assay for the quantity of a specific body chemical constituent which may be abnormally increased or decreased in amount by the primary absence of a crucial enzyme.

Screening for multifactorial disorders, in some cases, also depends upon the detection of abnormal levels of certain body constituents. In other cases, early manifestations of disease are sought by physical examination.

Chromosomal disorders can be screened for by one of two methods. Usually, karyotypes are prepared, but sometimes, examinations for either or both male and female sex chromatin are performed.

Screening for Single-Gene Defects

Autosomal Recessive Disorders

The survey of populations for autosomal recessive inborn errors of metabolism has been the form of genetic screening most prominently in the public attention and mind. By far the greatest part of screening resources to date have been expended on three diseases: phenylketonuria, where the goal is to diagnose the disease early enough to permit effective therapy; Tay-Sachs disease, where the attempt is made to identify healthy heterozygous carriers, so they may have the opportunity to avoid having affected children; and sickle cell anemia, where both homozygous affected and heterozygous carrier persons are identified.

Screening for early diagnosis of disease: Phenylketonuria (PKU). Phenylketonuria is an inborn error of metabolism. About 1:10,000 newborn infants in the United States is an affected homozygote, and one in fifty persons is a heterozygous gene carrier. The condition is unusual in blacks and is most common in persons of Northern European extraction (alas, the poor eugenicists).

Phenylketonuria was first identified in 1934, by the Norwegian physician Ashborn Folling. A mother had brought Dr. Folling her two retarded children because they had an unusual odor. In the ensuing workup, it was discovered that when ferric chloride was mixed with the urine on the children's diapers, a blue color appeared. The substance in the urine responsible for this reaction was identified as phenylpyruvic acid, an unusual metabolic product of the amino acid phenylalanine.

Chemically speaking, phenylpyruvic acid belongs to a class of compounds known as phenyl-ketones (Fig. 3.1): hence, the name phenyl-keton-uria was given to the disease. Another name that has sometimes been used is phenylpyruvic oligophrenia ("oligophrenia"

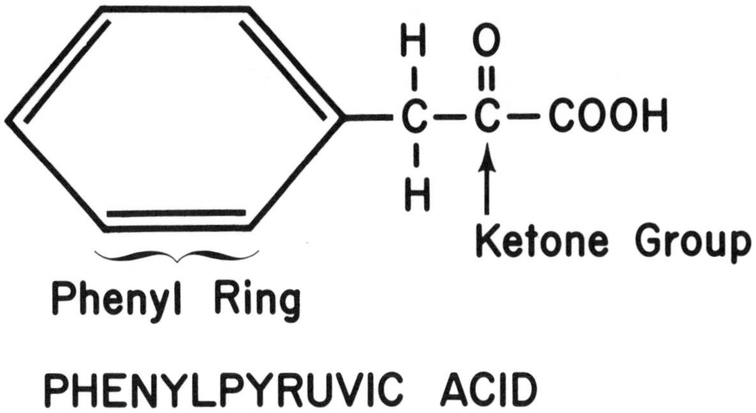

H   O
| ||
—C—C—COOH
| ↑
H   **Ketone Group**

Phenyl Ring

PHENYLPYRUVIC ACID

Fig. 3.1: The chemical structure of phenylpyruvic acid.

meaning "little mind"). In addition to severe mental retardation, patients with PKU may have seizures and eczematous skin lesions. Commonly, they are lightly pigmented, with blue eyes, blond hair, and fair skin. It has been determined that about one in every hundred inmates of mental institutions suffers from PKU.

After Dr. Folling's discovery, it was determined that the basic defect in PKU is absence of phenylalanine hydroxylase, one of the enzymes necessary to convert phenylalanine to another amino acid, tyrosine. This biochemical deficiency is responsible for all the clinical features of PKU. The abnormally high levels of phenylalanine which accumulate in the brain prevent other necessary nutrients from reaching the cells; at the same time, since normal metabolic transformation of phenylalanine is impossible, this amino acid is degraded to abnormal compounds such as phenylpyruvic acid. The generally deranged biochemical environment results in malnourishment of the brain cells, with consequent mental retardation and seizures. The skin reacts to the aberrant metabolism with eczema; furthermore, since tyrosine, the amino acid that normally should be formed from phenylalanine, is necessary for skin and hair coloration, the metabolic block accounts for the generalized underpigmentation. And the phenylalanine and phenylpyruvic acid in the urine cause the characteristic musty odor.

After the biochemical disturbances of PKU had been elucidated, it was discovered that a special diet based on protein from which the phenylalanine had been extracted had a general salutary effect on institutionalized phenylketonurics, improving their behavior and reducing their frequency of seizures. However, the degree of mental retardation remained unchanged. Therefore, an attempt was made to prevent the occurrence of mental retardation by feeding the low-phenylalanine diet to affected infants shortly after they were born. It was reasonable to believe that PKU infants would be normal at birth, since during intrauterine life, maternal phenylalanine hydroxylase can circulate via the placenta to the fetus, thereby maintaining the fetal phenylalanine concentration at normal levels.

But how to identify the newborn baby with PKU? Until the high concentrations of phenylalanine have done their irreversible damage, these children look like any others. So, Dr. Folling's ferric chloride test was put into use, and with that, newborn screening for PKU began. A few drops of ferric chloride solution were placed on a wet diaper from each baby, and a blue-green color was looked for.

This was a start, but it wasn't altogether satisfactory. For one thing, phenylpyruvic acid is not excreted in high levels in the urine until large quantities of its precursor, phenylalanine, have existed in the blood for about a month. Therefore, by this time, there already may have been significant brain damage. In addition, it is not easy to track down all month-old children, so too many babies never were screened. Another problem is that phenylpyruvic acid is not stable, so unless the urine is tested soon after passage, the typical color may fail to form, leading to a false-negative result.

This, then, was the situation in 1961, when Dr. Robert Guthrie introduced his now-famous blood test for the screening of PKU in newborn infants. A drop of blood is obtained from the heel of the baby, absorbed in a small piece of filter paper, and sent to a testing center. There, it is placed on a plastic plate whose surface is covered with a coating of gel containing small numbers of a microorganism known as Bacillus subtilis. B. subtilis normally grows nicely on gel, but Dr. Guthrie's gel preparation is also impregnated with β-2-thienyl alanine, a compound which prevents B. subtilis from synthesizing the phenylalanine necessary to its growth and reproduction. However, if a blood sample happens to contain a large quantity of phenylalanine, this will diffuse from the filter paper into the gel and permit growth of the B. subtilis. A positive test, then, constitutes a ring of bacterial colony growth around a piece of filter paper.

Logically enough, this procedure is called a *bacterial inhibition assay*. The plastic plates are large, so that there is room on each one for pieces of filter paper from between fifty and 100 infants. When a positive or a suspicious test is seen, the infant is called in for a direct blood-serum assay of phenylalanine. If this reveals circulating phenyl-alanine levels in excess of those considered safe, the infant is presumed to have PKU, the situation is explained to the parents, and the child is started on the low-phenylalanine diet. Basically, the diet consists of a formula containing milk whose casein (milk protein) has been hydro-lyzed, or broken down, into its constituent amino acids, thereby permitting removal of the phenylalanine. As the child gets older, the casein hydrolysate continues to be his major source of protein; only small quantities of meat, cheese, or beans are allowed. The formula is supplemented with vegetables, fruit, and fats.

The low-phenylalanine diet is not palatable, due to the unpleasant odor of the casein hydrolysate. However, it was found to be capable of preventing mental retardation when begun within two or three weeks of birth. Further studies showed that the diet could be discontinued at about age six without harmful consequences: presumably, the critical period is the time of rapid brain growth during the first few years of life.

At this point, newborn screening for PKU seemed to have arrived at the millenium. Hospitalized babies constituted an easily accessible target population. The test used was simple, reliable, inexpensive, and easily done in mass lots. Follow-up was straightforward, and there was a reasonable mode of therapy. The screening program appeared so promising, in fact, that doctors and involved laymen pressed for action, and legislators responded. Forty-three states passed laws making PKU screening mandatory in newborn infants, and in the other states, widespread testing was done on a voluntary basis. Altogether, by the late 1960s, about 90 percent of all babies were being screened.

As time passed, however, it became apparent that PKU screening was not quite as free of problems as had been thought. Coincident with the trend toward earlier hospital discharge of mothers and babies, an increase was noted in new cases of symptomatic PKU. This was found to be due to the fact that the elevation of blood phenylalanine levels in affected babies may not become apparent until the third day of life or later. (Remember, babies with PKU start off with normal levels of phenylalanine, due to "borrowed" maternal phenylalanine hydroxylase.)

In addition, analysis of positive Guthrie tests made it obvious that there was a significant excess of affected males. However, surveys of institutions for the retarded had always shown equal numbers of affected boys and girls; in addition, pedigrees of families of PKU victims were thoroughly consistent with autosomal recessive inheritance. The puzzle was solved by the discovery that girls with PKU have a slower initial rise in blood phenylalanine levels than do affected boys. Therefore, more affected girls were being overlooked by the Guthrie test.

A serious problem was the finding that all infants with increased circulating phenylalanine do not have PKU. Some children have mild or moderate elevations in their blood phenylalanine levels which are not associated with any physical or mental disability. This benign variant is probably caused by an allele which codes for a partially active phenylalanine hydroxylase enzyme molecule. Tragically, the low phenylalanine diets given to these children led in some cases to abnormally low levels of body phenylalanine; in some cases, this was fatal.

The doctors in charge of the screening programs moved promptly to overcome these problems. Arrangements were made to test all babies no earlier than the third day of life. The bacterial inhibition assay was refined so as to provide a more accurate estimation of the blood level of phenylalanine. In this way, babies who showed only a slight increase over the normal when tested between the third and the eighth day of life could be identified for retesting. Special attention was paid to girls with minimally elevated readings, formerly thought to be insignificant. Furthermore, follow-up procedures were greatly tightened, including the performance of more frequent assays of blood phenylalanine levels in children on diets; this was done in order to prevent possible overtreatment. Greater experience and careful study helped the doctors to define levels of blood phenylalanine which could be considered definitely normal, definitely benign elevation, definitely PKU, or equivocal. In this way, the rigors of the special diet could be avoided for many of the children discovered to have elevated blood phenylalanine levels.

Mass screening of newborns for early diagnosis and subsequent dietary therapy has also been found to be possible for several inborn errors of metabolism which occur less frequently than PKU. The best known of these conditions is galactosemia, the inability to metabolize galactose (milk sugar). A galactose-free diet will spare affected children the mental retardation and physical defects characteristic of this disease.

At this time, all the problems associated with routine newborn screening for PKU and other inborn errors have not been solved. However, the programs generally work well and seem worthwhile. Certainly, they have given many people the priceless gift of mental normality, saving them from hopeless lives in mental institutions.

Screening for detection of healthy gene carriers: Tay-Sachs disease (TSD). This dreadful disorder is named for the British ophthalmologist Warren Tay and the American neurologist Bernard Sachs, who were the first physicians to study and describe the clinical features of the disease. For the first half-year of their lives, babies with Tay-Sachs disease appear to be normal, healthy infants, developing properly in all ways. But then, they gradually deteriorate, becoming weak, listless, and blind. Next, spasticity and total dementia develop. Seizures occur with increasing frequency. The condition is invariably fatal, with death usually occurring between the ages of three and four.

There is no treatment for Tay-Sachs disease, and the financial cost of caring for a baby with this problem can run as high as $50,000 a year. The emotional cost cannot be imagined by someone who has never been involved in such a situation. Small wonder that until recently, couples who had had a Tay-Sachs baby usually did not attempt another pregnancy, with the attendant 25 percent recurrence risk.

Thanks to pedigree studies, it has long been apparent that Tay-Sachs disease, like PKU, is inherited by an autosomal recessive mechanism. Also like PKU, TSD is an inborn error of metabolism. Tay-Sachs victims are homozygotes for a gene mutation which results in absence of an enzyme known as Hexoseaminidase A, commonly shortened to Hex A. Hex A helps to metabolize a lipid called ganglioside GM2, which is a normal constituent of brain tissue. In the absence of Hex A, the brain cells accumulate huge quantities of ganglioside GM2, which, like phenylalanine in PKU, severely disturbs normal cellular vital processes, leading to clinical disease.

Phenylalanine hydroxylase is a liver enzyme, but Hex A is found in many body cells, including the fetal cells in amniotic fluid. Therefore, as will be explained in chapter 5, TSD can be diagnosed in the early fetus by performing Hex A assays on cultivated amniotic fluid cells.

The gene for TSD occurs with by far the greatest frequency in Ashkenazi (Eastern European) Jews. One in twenty-five Jews of Eastern European origin carries the Tay-Sachs gene, and the incidence of TSD in this population is one in about 3,000 newborns. In contrast, the carrier frequency in Gentiles and non-Ashkenazi Jews is one in 300; the rate of TSD among these newborns is only one in 360,000.

Newborn screening for an incurable disease would be pointless. But the facts described in the two preceding paragraphs gave Dr. Michael Kaback an idea. Maybe a different sort of screening would be useful in combating TSD.

In 1971, Dr. Kaback was a pediatric geneticist at Baltimore's Johns Hopkins School of Medicine. Over a period of approximately six months, he and his colleagues performed a tremendous feat of mass genetic counseling. Directing their information toward the Jewish population of Baltimore and Washington, D.C., Kaback's group disseminated information concerning the inheritance and clinical nature of TSD, and how the disease could be prevented. Working closely with the community's religious leaders, the geneticists tried to make certain that their prospective screenees fully understood the proposed program. Only then did they begin to screen, setting up centers for the obtaining of blood samples and the providing of necessary follow-up information to individuals.

The primary test for heterozygous Tay-Sachs carriers is based upon the fact that the quantity of Hex A in the circulation is rather well correlated with the number of functional genes for the production of this enzyme possessed by the individual. A homozygous TSD patient has no circulating Hex A, and a heterozygous carrier has approximately half the concentration of someone with two functioning Hex A genes. Therefore, Dr. Kaback's group accepted Jewish men and women of childbearing age, drew a blood-serum sample from each, and performed mass chemical assays on the samples for Hex A concentration. The screening was not limited to Jews of Eastern European background, because so many Americans are not certain about their ancestry, geographically speaking.

Whenever a person was found to have a serum Hex A concentration within the carrier range, he was recalled for a second test. This time, Hex A assays were performed on drawn samples of white blood cells. Although the white-cell assay is too complex for routine screening use, it is less likely to give false-positive results. Thus, it is ideal for a confirmatory follow-up test. The white blood cell assay was also employed when the results of the serum assay were not frankly in the carrier range, but were suspicious. Also, since pregnant women and women taking birth control medication are likely to have false-positive serum tests, white blood cell assays were used for these persons.

Whenever a person was determined to be a carrier, he was called back for further counseling. When only one member of a married couple carried the gene, it was explained that there would be no problem with TSD in the children. However, if both husband and wife proved to be heterozygous, Kaback's team carefully explained the nature of autosomal recessive inheritance, including the fact that the odds for a Tay-Sachs child were 25 percent for each pregnancy. Then, a description was given as to how the couple could avoid having such a child, using prenatal diagnosis.

Notice the difference between the descriptions of screening for PKU and TSD. The former is an example of generalized screening: the goal is to test every possible newborn baby for PKU. The latter, however, exemplifies selective screening, where a population at especially high risk is identified and examined. More will be said later about generalized vs. selective screening.

Was the Baltimore-Washington Tay-Sachs survey worth it? Approximately 10,000 persons were screened, about 400 carriers were identified, and eleven couples were found where both husband and wife carried the gene. The cost of care for the number of Tay-Sachs children expected from these eleven matings would be about five times the cost of the entire screening program. This, of course, is in addition to the unspeakable mental anguish spared the couples at risk.

Screening for both affected persons and gene carriers: sickle cell anemia (SCA). Hemoglobin, the oxygen-carrying protein of the red blood cells, is a relatively large molecule. Thus, it would be expected that the gene which codes for its production might be a relatively long region of DNA, and therefore subject to many possible mutations, each of which would produce its own characteristic variation in structure of the hemoglobin protein molecule. This is indeed the case, and in fact, hundreds of hemoglobin variants have been identified. Most of these are mere medical curiosities, in that they function as well as does "normal" hemoglobin. Such is not true, however, of hemoglobin S, the abnormal protein which causes the disease of sickle cell anemia.

In the presence of low oxygen concentrations, the morphology of hemoglobin S becomes abnormal, which causes the red blood cells containing it to assume a sickle shape, rather than the normal "doughnut" shape. Abnormally shaped red cells are more likely to be destroyed in the body; this overdestruction produces the anemia. Furthermore, the sickled cells tend to "stick" in small blood vessels, blocking the vessels and producing clots in such tissues as bones, lungs, and spleen. The blockage of the blood vessels prevents oxygen from reaching the area; this in turn worsens the sickling, and so sets up a vicious cycle. Such situations characterize the painful "sickle crises," which may culminate in death. Many patients with SCA do not survive their early twenties, but others may reach middle age.

SCA is inherited by an autosomal recessive mechanism. Persons with two nonmutated hemoglobin genes produce the normal hemoglobin A; those with two "sickle alleles" produce hemoglobin S. Heterozygous carriers produce both hemoglobins A and S, and so, their red blood cells contain a mixture of the two hemoglobins. Since hemoglobin A has a strong protective effect on the cells with regard to sickling, heterozygotes have normal expectancies for life and health. Only rarely do they show clinical sickle manifestations, and then it is under conditions of severely reduced oxygenation, such as high-altitude flying in unpressurized aircraft.

The gene for sickle hemoglobin occurs with especially high frequency in blacks. In American blacks, approximately one in twelve persons is a carrier, and about one black baby in 600 is homozygous and thus has SCA. So, there is a high rate of both carriership and disease in a well-delineated subpopulation. In addition, there are several simple, reliable tests to identify hemoglobin S in small samples of blood. When a positive test for S-hemoglobin is found, it then must be determined whether the subject is a homozygote or a heterozygote for the S-mutation. For this *hemoglobin electrophoresis* is used: a drop of blood is placed at the end of a cellulose acetate column, and an electric current is applied to the column. Hemoglobins of different composition will migrate through the current at different rates, so that the blood from a homozygote for either hemoglobin A or S will produce a single spot at the location on the column characteristic for that hemoglobin. However, the hemoglobins A and S of a heterozygous carrier will separate, producing two spots (Fig. 3.2).

Considering the information in the above paragraph, it might be—and has been — supposed that sickle cell anemia offers a fine

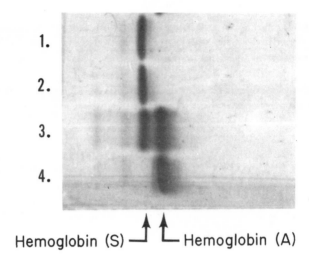

Hemoglobin (S) ⌐↑ ↑⌐ Hemoglobin (A)

1. & 2. Sickle Cell Anemia
(Homozygous Hemoglobin S)

3. Sickle Trait Carrier
(Heterozygous Hemoglobin A-S)

4. Normal (Homozygous
Hemoglobin A)

Fig. 3.2: **Hemoglobin electrophoresis: (a) homozygous hemoglobin A (b) homozygous hemoglobin S (c) heterozygous hemoglobin A-S. (Courtesy Dr. James Detter, University of Washington School of Medicine.)**

opportunity to screen for both carriers and disease. But there's one problem. There exists no truly effective therapy for SCA, nor is there a prenatal diagnostic technique whereby the disease can be prevented. At this time, then, screening for hemoglobin S can provide little in the way of practical, usable information.

Unfortunately, however, the advent of screening techniques happened to coincide in time with the rising tide of "black awareness" in America. Thus, sickle-cell screening seemed to offer some well-meaning persons the opportunity to "do something" in the way of giving attention to the health problems of blacks. But as is the case with most hastily conceived plans, the results have not been good.

Sickle-cell screening programs linked to follow-up counseling produced a wave of so-whatism among those diagnosed as carriers, whether or not they were also found to have affected children. In the latter situation, no treatment could be offered the children. Carriers without affected offspring fared little better in terms of useful information. If married to another carrier, they have only one alternative to taking the one in four risk of disease associated with each pregnancy — not to reproduce at all, an unsatisfactory choice. Unmarried carriers were told that they might check out the genetic makeup of any member of the opposite sex who was beginning to look attractive. This suggestion met with a predictable lack of enthusiasm. In the end, then, the screenees paid a high price in anxiety, for which they received little in the way of return.

To make matters worse, some members of the black lay community, deciding that "black problems" could be best dealt with by blacks, began loosely organized door-to-door sickle-cell screening drives. Some groups were to a degree affiliated with established genetic counseling services, but others went it completely alone. Too often, the screening procedure consisted of ringing a doorbell, drawing a sample of blood from each inhabitant of the house, testing the bloods on the spot for sickling, and saying something to the effect of, "You and little Lucy are OK, Mrs. Jones, but little Johnny has 'sickle cell.'" Then, the screeners would proceed to the next house, leaving a frantic Mrs. Jones with the vague idea that her child suffered from some sort of terrible disease. She couldn't know — nor could anyone, without doing a hemoglobin electrophoresis — whether Johnny's positive screening test represented the usually harmless heterozygous carrier state, or the more serious homozygous SCA. In addition, without antecedent counseling, it would not have been likely that she even would have known of the existence or the meaning of the carrier state.

As might be expected, one result of the now decreasing interest in sickle-cell screening has been a rash of "sickle-cell neurosis" among healthy persons diagnosed as carriers. The suspicion persists among many persons, black and white, that the heterozygous state represents a milder but still serious form of the disease.

Thus, we can see the effect of screening without proper counseling, in a situation where no use can be made of the information obtained. When sickle cell disease can be effectively treated and/or prevented (as by prenatal diagnosis), sickle screening of blacks will be a reasonable procedure. But not before that time.

Autosomal Dominant Disorders

Coronary heart disease is the leading cause of death in America, and it is generally assumed that persons with hyperlipidemia (elevation of certain fats, or lipids, in the blood) are at increased risk of heart attack. Furthermore, it has been estimated that as many as one percent of the population have a familial type of hyperlipidemia which seems to be inherited as an autosomal dominant trait. Combination drug and dietary therapy appears to lower the blood lipids in these patients.

Many questions remain to be answered: Does this therapy also lower the incidence of heart attacks? Is the therapy safe as well as efficient? At what age should the therapy be started? When we have the answers to these questions, it might then make sense to screen newborns for serum lipid levels. This might be done in a generalized fashion, or attention might be paid to the offspring of parents previously ascertained to have elevated blood lipids. In such a way, we could conceivably achieve our first true breakthrough in the prevention of arteriosclerotic heart disease.

No other autosomal dominant diseases seem to offer opportunities for screening in the foreseeable future. Most of these conditions are rare, and can be neither treated nor prevented.

X-linked Recessive Disorders

Glucose-6-phosphate dehydrogenase (G6PD) is an enzyme that plays a part in the metabolism of glucose. It is coded for by a gene located on the X chromosome. Mutations which produce deficiency of G6PD are common among both whites of Mediterranean ancestry, and blacks. About 12 percent of black males carry a defective G6PD gene; since males are hemizygous for X-linked genes, this is the only G6PD gene they have. Ordinarily, this situation does not interfere with health, but in some unknown fashion, ingestion of certain drugs can trigger a wave of hemolysis (red blood cell breakdown) in G6PD-deficient individuals. This results in severe anemia, often associated with racking abdominal pain. Most of the drugs involved are uncommonly used, but the list does include the ubiquitous sulfas. G6PD activity can be determined by a simple and reliable blood test. Therefore, screening has been suggested for ethnic groups at high risk for G6PD deficiency, so that persons with this condition may avoid dangerous drug exposure. In addition, should they happen to suffer an attack, they would be in the position of being able to alert their medical attendants as to the reason for their symptoms.

Aside from G6PD deficiency, most of the attention regarding screening for X-linked conditions involves attempts to identify (female) carriers for such conditions as hemophilia and Duchenne muscular dystrophy. No specific population has been identified as being at special risk for either condition, and no simple or reliable test has been devised in either case. However, to demonstrate an important basic principle, I'll briefly describe one of the imperfect procedures that may be useful in selected cases to detect carriers for Duchenne muscular dystrophy.

Remember that in the Introduction, I stated that there exists one minor difference between the carrier of an autosomal recessive disease and the carrier of an X-linked recessive trait? That difference is at the cellular level. Since each cell contains one full set of chromosomes (and of genes), each cell in the body of an autosomal recessive carrier contains one functional (and one nonfunctional) gene for the trait in question. However, since there is random inactivation of one of the two X chromosomes in each cell of a female, this leaves each cell of an X-linked trait carrier with *either* a functional *or* a nonfunctional gene. Thus every cell of an autosomal recessive carrier is in proper working order, but this is true of only half the cells of X-linked carriers. This curious fact underlies the test currently being used to try to identify the carrier status of female relatives of patients with Duchenne muscular dystrophy.

Creatine phosphokinase (CPK) is an enzyme active in muscle metabolism. It leaks into the circulation from the damaged muscles of boys with muscular dystrophy, so that the circulating levels of CPK in patients suffering from this disease are as high as one hundred times normal. Since about half the muscle cells of a carrier woman contain only the defective "Duchenne gene," these individual cells sometimes leak enough CPK to raise the level of this enzyme in the blood to two to three times normal. CPK levels are elevated in about two out of three *obligate carriers:* women shown to be carriers by pedigree analysis (Fig. 3.3). Thus, a woman whose affected son represents the first case in her family is proved to be a carrier if her blood CPK level is elevated. However, if her CPK is normal, she still may be a carrier; alternatively, her son's case may represent a new mutation. CPK levels cannot be used for general screening, because a reliability level of two-thirds is not nearly sufficient. In addition, the procedure is not a simple assay.

Generally speaking, it would seem that screening for carriers of X-linked diseases is not to be expected in the near future.

Screening for Multifactorial Disorders

Many localized screening programs have been initiated for the early diagnosis of two multifactorial disorders, diabetes mellitus and hypertension (high blood pressure). These programs have suffered from common shortcomings. First, there exists considerable doubt as to the diagnostic reliability of single blood pressure readings and single blood sugar levels. Second, counseling and follow-up usually have not been well coordinated, so that considerable patient anxiety has been reported. Third, there is great disagreement regarding the universal desirability and efficacy of treatment in the presence of positive tests. Therefore, I think that until the basic four premises outlined at the

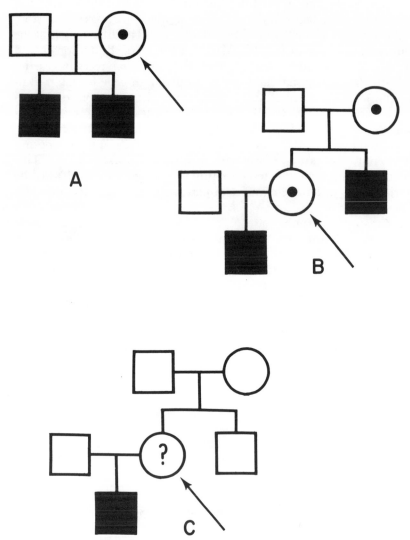

Fig. 3.3:   Patients with genetic diseases can represent the effects of either
"old" transmitted mutations, or "new" fresh mutations. The indicated women in (a) and (b) are assumed to carry the gene for
muscular dystrophy, because it is extraordinarily unlikely that a
fresh mutation for the same trait would occur in two sons, or in a
son and a brother. However, as illustrated in (c), a single case in a
family may represent either a transmitted mutation (with a recurrence risk of 50 percent in sons) or a fresh mutation (with a
negligible recurrence risk). Maternal CPK levels can sometimes
resolve this dilemma.

beginning of this chapter can be satisfied, it might be better to leave testing of blood pressure and blood sugar to the discretion of individual doctors and patients, as part of routine annual checkups.

One multifactorial situation for which general screening may soon be appropriate is the anencephaly/spina bifida complex of nervous system birth defects. These conditions, described in chapter 5, are associated with high amniotic-fluid levels of a compound known as alpha-fetoprotein. Since alpha-fetoprotein has also been identified in abnormal concentrations in the blood of some women pregnant with fetuses afflicted with anencephaly or spina bifida, it is hoped that this may lead to the development of a routine blood analysis for the prenatal recognition of these common and very serious birth defects.

Screening for Chromosomal Disorders

The last ten years have produced a large number of reports of chromosomal screening programs. Some of these have been generalized surveys of newborn infants, children, and adults. Others have been selective: studies of populations composed of prisoners, inmates of mental institutions, or older mothers. These programs have uncovered both disease-producing unbalanced karyotypes, and balanced variants (such as translocations), which are capable of causing problems in the next generation.

Chromosomal screening programs have been invaluable in the advancement of genetical knowledge. Because of them, we now know that one in every 200 infants is born with a significant chromosomal defect. We know the types of mental and physical aberrations that are typically associated with different chromosomal anomalies. We have also learned that most patients with XYY and XXX are normal, but that a larger number of them than would be expected show psychological and behavioral abnormalities.

This gathering of information is not the plaything of head-in-the-clouds scientists. Chromosomal diseases are relatively new to our recognition, and the first step in dealing with them is to delineate precisely which clinical abnormalities are associated with which chromosomal abnormalities. We also must learn the frequencies of occurrence of the different conditions. Such data will provide us the basis by which we may intelligently order our future health-care priorities.

Furthermore, it is not true that chromosomal screening is never useful to the individual patient. For example, there is suggestive evidence that boys with Klinefelter syndrome (XXY) may suffer less psychopathy if testosterone (male sex hormone) therapy is begun at about the time of puberty. In addition, girls with Turner syndrome

(XO) usually do better emotionally if estrogen (female sex hormone) replacement treatment is begun before they have had to suffer through a number of years of sexual infantilism and lack of menstrual periods, while their friends have been blossoming all around them. Without newborn or childhood chromosomal screening, neither Klinefelter nor Turner syndrome would often be diagnosed prior to the late teens. Other persons who probably would appreciate chromosomal screening are the carriers of balanced translocations since they may then apply their information by deciding to utilize prenatal chromosomal diagnosis, in order to avoid having children with unbalanced karyotypes.

Thus, chromosomal screening is a good illustration of the manner in which medical research can interfuse with medical care, advancing our general level of knowledge as well as providing useful information to patients. Such programs as those currently under way at many of our major universities would seem to deserve enthusiastic support.

At this time, the major hindrance to the wide utilization of chromosomal screening is the unavailability of mass-analysis techniques. One possibility for the future is the computer chromosomal analysis being developed by scientists at the Jet Propulsion Lab at Cal Tech. It now takes a technician a good day's work to analyze a karyotype for only one patient, but computers should soon be able to provide accurate readings on hundreds of patients a day. This would be necessary to handle the load, should generalized population screening become routine.

Moreover, it should be emphasized that chromosomal screening includes the responsibility of careful, accurate counseling of those found to carry abnormalities, including long-term follow-up to ensure the availability of appropriate therapeutic and/or preventive measures, as well as to provide counselees updated information as new data become available.

Some Controversial Aspects and Future
Directions of Screening

Whip vs. Stick and Carrot

Much of the argument advanced in support of compulsory screening is based upon the concept of genetic disease as "communicable," and therefore representing a threat to society. Another argument in favor of compulsory screening is that the state is obligated to act in the interests of the helpless — the newborn babies. To my mind, neither of these premises is reasonable.

With traditional communicable diseases (infections), the great hazard is transmission of disease "horizontally," that is, to other living persons. With genetic diseases, however, the problem is one of "vertical transmission" of the disorders to future generations. It seems a tenuous concept to consider as communicable those diseases that can be transmitted only to theoretical persons. Making the situation even more difficult is the fact that the illnesses can be prevented only by preventing the "victims" from ever being born. While I would whole-heartedly support the prerogative of parents to abort diseased fetuses, I do not try to justify this approach by deciding that the fetus would prefer no life to impaired life. Furthermore, it's not likely that all prospective parents forced to undergo screening would voluntarily utilize their information. So, the compulsory prevention of genetic disease cannot logically end with the screening process. How far would we want to go?

Compulsory genetic screening of the newborn is not consistent with many of our other attitudes. Parents whose religious or philosoph-ical beliefs preclude vaccination are not forced to have their children immunized. Despite the increased risks to the baby of home delivery, a fair number of women are currently giving birth in this fashion. Why should genetic disease be considered in a different light than these and many similar situations?

Actually, the entire argument for compulsory screening becomes pointless when we consider data such as that from Kaback's Tay-Sachs screening programs. This information demonstrates nicely that after appropriate education, the large majority of the target population elected to be screened. Furthermore, those declining the testing were mostly those who could see no personal use for the tests: couples who had completed their families, or couples who would not consider abortion of an affected fetus. Not only is this appropriate, but it also points up the absurdity of battling over the "rights" to know and not to know. Under voluntary plans, people receive information that permits them to decide for themselves whether they will be better off knowing or not knowing.

Should Screening Be Generalized or Selective?

Generalized screening is only feasible when the risk of the disease in the general population is high. Thus, it becomes impractical, financially or otherwise, to screen for inborn errors of metabolism that are much rarer than PKU (about 1:10,000). However, combined simul-taneous screening of one blood or urine sample for several diseases would make more sense. Dr. Guthrie of PKU fame has devised modifications of his test to permit simultaneous screening for a number of rare metabolic diseases. In addition, some labs, primarily in Scandi-navia, are developing techniques of electrophoresis and spectral analy-sis that will permit screening of one blood or urine specimen for over a thousand different compounds. If and when such techniques become routine, they may offer the opportunity to test large populations for general metabolic profiles. An abnormal level of any compound found would mandate specific investigative follow-up.

Selective screening, however, aimed at high-risk populations, probably presents the best opportunities to diagnose genetic diseases and to detect asymptomatic carriers. In addition, the smaller and more homogeneous populations probably would permit more satisfactory counseling and follow-up. But, because selectivity tends to isolate ethnic groups, there can be trouble. For example, the attention paid to the detection of sickle cell disease has caused some blacks to worry about possible genocidal attempts. On the contrary, after the District of Columbia had screened almost 80,000 (predominantly black) newborns for PKU without finding a single case, the cancellation of further testing caused other black leaders to protest the action as discriminatory. I suppose that such problems can best be avoided by making screening totally voluntary: no one has to be screened who doesn't want to be, but, in addition, screening for a particular disease should not be denied to someone who wants it, even though such a person might not be a member of the high-risk population.

Risks of Undergoing Genetic Screening

Many of the problems of genetic screening have been discussed in chapter 1. They include psychological difficulties caused by the knowledge that one carries a genetic disease, social stigmatization, and increased insurance policies. There is real concern regarding occupational discrimination, exemplified by the refusal of some airlines to hire sickle cell heterozygotes as stewardesses.

All of these problems are genuine, but they do not constitute a rationale for the prohibition of genetic screening, when the results would permit therapeutic or preventive options. No one likes to get the bad news that he has or carries a disease, be it genetic, infectious, or cancerous. But in any case, the disease will still be there, and it would seem advantageous to avail oneself of all possible information, in order to deal with the situation in as reasonable a manner as possible. Education probably would be of great help in this regard, as it would also be for the problem of social stigmatization. Education should also settle the problem of job discrimination, once people become familiar with the true nature of different genetic diseases.

Not so many years ago, epileptics were routinely classed with the "mental defectives"; it took a good deal of hard work to convince most of the population that the great majority of epileptics can lead normal lives. As to the question of higher insurance rates, this unfortunate situation can be the result of the discovery of any disease in a person. Should we stop diagnosing all diseases, so as to keep insurance rates uniformly low?

New Problems for Old

As is true in most other circumstances, the solving of one problem usually leads to the creation of a new one. Before the dietary therapy of PKU became a reality, the severely retarded victims of this disease never reproduced. But, as the first batch of treated phenylketonurics grew to adulthood, a disturbing fact was observed: a high proportion of the offspring of phenylketonuric women were severely mentally retarded. This was not consistent with autosomal recessive inheritance. In addition, most of these babies suffered from microcephaly (small heads), a feature not characteristic of PKU. To make the story short, it was discovered that the abnormalities in the babies were caused by the high maternal levels of phenylalanine. (Remember, the special low-phenylalanine diet can safely be stopped at age six.) Large circulating quantities of phenylalanine, harmless to the adult, cross over to the fetus and produce irreparable damage to the fetal brain. Therefore, it is now considered imperative to maintain phenylketonuric women on the low-phenylalanine diet from the earliest time in pregnancy. They must be warned long in advance that this will be necessary, and that it would probably be best were they to reinstitute the diet prior to planning a pregnancy. Since the diet is unpalatable, the utmost efforts must be expended toward motivation; the patients must be told graphically what will be the result for their babies, should they fail to follow dietary instructions.

Genetic screening has been criticized because "too much attention is paid to the detection of rare diseases." This objection is inappropriate. Both economically and emotionally, it has been repeatedly demonstrated that it is cheaper to detect and then treat or prevent these genetic diseases than to provide lifelong care to the crippled and retarded victims. As satisfactory means of treatments and prevention become available for more diseases, and as multiphasic and specific screening techniques are refined, we may expect genetic screening to assume an ever-increasing role in assuring better health to us and our children.

4

Genetic Therapy

During the first half of the twentieth century, by far the greatest share of medical attention was focused upon infectious diseases. With the eventual conquest of this group of killers, doctors began to look for new challenges. They didn't have to look far. The advent of the antibiotic era coincided with research breakthroughs in cell biology (including the elucidation of the structure of DNA), chromosomology, and biochemistry; all these advances contributed greatly to our understanding of hitherto mysterious inherited diseases. Concomitantly, the major medical and surgical advances of the time were rescuing many victims of genetic diseases from the early deaths which had previously been their universal fate. For example, a baby would die of pneumonia, with no one recognizing that its lungs had only become infected because it was a victim of cystic fibrosis.

In spite of this progress, there has been persistence of the hopelessness and negativism traditionally associated with genetic diseases. Even today, the general reaction is that we can now give names to these conditions but we still can't cure them.

However, this general pessimism is not appropriate. Understanding of mechanisms of disease is a necessary prerequisite to satisfactory therapy. Furthermore, within the past decade, great strides have indeed been made in the prevention and treatment of genetic diseases. Prevention will be discussed in chapters 5 through 10. In this chapter, we'll consider the methods currently available for the treatment of genetic diseases, and speculate as to what kinds of genetic therapies may appear in the future.

A broad interpretation will be applied to "genetic therapy": the term will be taken to mean any manner of treatment whose purpose it is to forestall or mitigate the harmful effects of genetic abnormalities upon an individual. In relation to the point in the disease process at

LEVEL 3:

Defective DNA
Ultimate Source of Disease

LEVEL 2:

Abnormal Gene Product
Intermediate Cause of Disease

LEVEL I:

Physical and/or Mental Symptoms
Manifestations of Disease

Fig. 4.1: The therapy of genetic disease may be attempted at three levels of complexity and sophistication. The third (highest) level of treatment involves correction of the defective gene itself; the intermediate level is represented by administration to the patient of a normal gene product; the first level of therapy is directly symptomatic.

which treatment is directed, genetic therapy can be attempted at three different levels (Fig. 4.1).

The first level of genetic therapy includes those techniques designed to deal directly with the signs and symptoms of disease. Such maneuvers represent compensatory rather than curative treatment, and include dietary manipulations, drug therapy, surgery, and avoidance of dangerous environmental agents.

The second level of genetic therapy consists of the administration of a normal gene product, such as an enzyme, to compensate for the patient's own defective or absent product. Not only is this a more complex scientific maneuver, but it also moves the point of application of therapy closer to the source of the disease and one step prior to the clinical manifestations. Thus, this level of therapy seems preferable to level one.

The third level of genetic therapy ought to be the most satisfactory of all: therapeutic manipulation of faulty DNA would present the opportunity to deal with the disease right at its origin. Correction of abnormal genes would constitute an actual cure, rather than the support and maintenance which are accomplished by therapy at levels one and two. Furthermore, transmission of disease-producing genes to future generations could be prevented by ameliorative alterations in the DNA of the sperm or the egg cells.

By far the greatest part of genetic therapy in current use is at level one; a few diseases can be dealt with at the second level, and there soon may be more. Third-level genetic treatment is presently at the stage of early research, and it is not likely that clinically useful techniques will be forthcoming in the near future.

As we consider the three levels of genetic therapy, we should keep in mind the major categories of genetic diseases: chromosomal, single-gene, and polygenic (multifactorial). At all treatment levels, both at the present time and in the future, the single-gene disorders offer the greatest opportunities for therapeutic intervention. This is logical, because the manifestations of chromosomal and multifactorial diseases are due to the combined and interacting effects of many genes. Just as it is easier to treat an infection due to one pathogenic bacterium than one due to many, it is easier to deal with a single malfunctioning gene than with several.

First-Level Genetic Therapy

Abnormal genes code for the production of nonfunctional proteins (often enzymes), which in turn play havoc with various aspects of cellular metabolism. These disturbances in body biochemistry then lead to the physical and mental aberrations which we recognize as signs

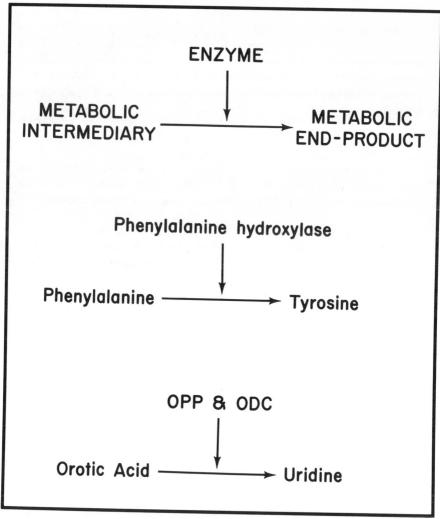

Fig. 4.2: An inborn error of metabolism results from deficiency of a
particular cellular enzyme necessary for the completion of a specific
metabolic reaction. Hence, the body accumulates an excess of the
intermediary compound that should be metabolized by the enzyme,
while there is a corresponding deficiency of the metabolic end
product. Symptoms of disease can be caused by either the excessive
intermediary or the lacking end product. In phenylketonuria (PKU),
mental retardation results from high levels of phenylalanine in the
brain; the deficiency of tyrosine seems to be harmless. On the other
hand, in orotic aciduria, the accumulated orotic acid is well tolerated, but the insufficient quantity of uridine precludes synthesis of
enough DNA to maintain health.

and symptoms of disease. Conceptually, the simplest modes of dealing with genetic diseases have as their aim either direct normalization of the deleterious metabolic processes, or the repair of physical defects. Special diets, drugs, surgical procedures, and avoidance of idiosyncratically dangerous substances all are useful first-level maneuvers.

Dietary Manipulations

The harmful effects of many gene mutations can be controlled and/or prevented by alterations in diet. There are about twenty-five genetic diseases for which dietary treatment has been attempted. In some cases, it has worked well, but in others, the results have not been nearly so satisfactory.

Dietary therapy finds its major application in the management of inborn errors of metabolism, most of which are inherited in an autosomal recessive fashion. Deficiency of a cellular metabolic enzyme causes accumulation of the substance that should be metabolized by that enzyme, as well as deficiency of the normal end product of the metabolic reaction. Signs and symptoms of disease may result from one or both of these two biochemical abnormalities (Fig. 4.2). Hence, the dietary therapy may be designed to reduce the quantity of accumulated intermediary compound, or to supply the end product which the patient is unable to synthesize.

The reduction of a harmful excess of metabolic intermediary is well illustrated by the dietary treatment of phenylketonuria (PKU), as described in chapter 3. Phenylketonuria results from deficiency of phenylalanine hydroxylase, an enzyme without which the amino acid phenylalanine cannot be metabolized to tyrosine; therefore, phenylalanine progressively accumulates in the body. A special diet low in phenylalanine prevents both this accumulation and the consequent derangement of cellular biochemistry, which, in turn, permanently damages the brain cells, producing severe mental retardation. Such treatment is satisfactory, since children maintained on the diet appear to develop perfectly normal intellectual capacities, and enjoy good general health. In addition, only the rapidly maturing brain of the young child seems to be susceptible to damage by high levels of phenylalanine, so that the unpalatable diet usually can be stopped at about the age of six.

Another disease which can be managed by dietary restriction of a metabolic intermediary is the autosomal recessive galactosemia. Lactose (milk sugar) is a compound carbohydrate which is broken down in the body to equal parts of the simple sugars glucose and galactose. Normally, the galactose is transformed metabolically to glucose, and then glucose is used in numerous energy-yielding body biochemical processes. However, in galactosemia, there is a deficiency of galactose-

1-phosphate uridyl transferase, one of the enzymes necessary to convert galactose to glucose. Hence, as is the case with phenylalanine in PKU, the unmetabolizable galactose rapidly builds up to toxic levels in many organs, causing mental retardation, cirrhosis of the liver, cataracts in the eyes, and usually childhood death. In most cases, all these problems can be prevented by a synthetic diet free of lactose and other galactose-containing compound sugars. Unfortunately, however, the body does not become less susceptible with age to the effects of excess galactose; consequently, it appears that galactosemic patients must be continued for life on the special diet.

Orotic aciduria is a good example of a condition where dietary treatment can supply a missing metabolic end product. This disease is a rare autosomal recessive whose major features are growth retardation, mental deficiency, and severe anemia. Illness results from the absence of two enzymes, orotidylic pyrophosphorylase (OPP) and orotidylic decarboxylase (ODC), which normally take part in the production of uridine, an important substance, since it is one of the components of nucleic acids. Since the function of OPP and ODC is to convert the metabolic intermediary orotic acid to uridine, the absence of these enzymes causes orotic acid to accumulate in the body. However, the situation is different from that in PKU and galactosemia, since excesses of orotic acid seem to be entirely harmless. It is the deficiency of uridine that causes ill health. Therefore, satisfactory treatment depends upon supplying the patient sufficient uridine, so his cells will be able to manufacture all the nucleic acids they need. This goal can be achieved by increasing the dietary intake of either free uridine or uridine-containing compounds. Once this is done, the patient is no longer compelled to produce his own uridine for nucleic acid synthesis, and so, the enzyme deficit becomes unimportant.

Drug Therapy

Several genetic diseases are now being treated effectively by specific drugs. Such disorders are exemplified by epilepsy, Klinefelter syndrome, and Wilson's disease.

Epilepsy is a condition characterized by repetitive fits, or convulsions. These seizures result from poorly understood periodic functional abnormalities in certain groups of cells in the brain. Some cases of epilepsy have their origins in brain-damaging head trauma; other cases are thought to be genetic, inherited by a polygenic mechanism. Throughout history, epileptics have been assigned widely variant social standings, ranging from their place of honor in the Roman Empire (no doubt due to the fact that Julius Caesar was an epileptic) to their consignment to asylums and forced sterilization procedures in the heyday of the American eugenicists. During the past quarter-century, the outlook for epileptics has improved markedly, thanks to drugs such as diphenylhydantoin and phenobarbital. These medications tend to normalize the brain dysfunctions which are responsible for epileptic seizures. In some cases, the drug effect is nothing less than dramatic: patients who had been totally incapacitated by daily seizures may go for months or even years without suffering a single convulsion. Thus, the vast majority of epileptics are now able to live normal lives, as long as they take their anticonvulsant medications faithfully. In addition, epileptics no longer need be concerned about the deterioration in intellectual functioning that may result from the frequent interference with oxygen supply to the brain consequent upon repetitive seizure activity.

As described in the Introduction, Klinefelter syndrome is caused by an extra X chromosome in males: these patients have 47 chromosomes, with an XXY sex chromosome complement. This condition is not rare; it is present in approximately one in 800 newborn males. The patients have small, sterile testicles, which produce little testosterone (male sex hormone). In addition, men with Klinefelter syndrome show a relatively high incidence of psychosocial aberrations, the most prominent of which are extreme passivity and poor capacity for interpersonal relationships. It has been suggested that these problems may be due in large part to low testosterone production, rather than to a direct effect on brain cell function by the genes on the extra chromosome. However, when injections of testosterone were given to grown men with Klinefelter syndrome, the resulting intensification of sex drive usually was more than these patients could handle; it was totally out of character with their long-established personalities. Therefore, as mentioned in chapter 3, some physicians are now giving testosterone in gradually increasing dosage to teenaged boys with Klinefelter syndrome, in order to reproduce the normal pubertal hormonal pattern. The results seem to be encouraging (if that term can be applied to normal adolescent behavior). Hopefully, when they reach adulthood, these treated patients will have fewer emotional difficulties than do untreated Klinefelter patients.

Wilson's disease is a rare autosomal recessive condition, characterized by the appearance during the childhood or adolescent years of progressive severe neuromuscular incoordination and spasticity, followed by cirrhosis of the liver. Untreated, the disease is invariably fatal. Although the specific enzyme responsible for the condition has not been identified, it is known that the clinical problems are due to inability to metabolize copper, so that toxic quantities of this metal are deposited in the brain, liver, and kidneys. It has become important to diagnose Wilson's disease before extensive pathologic changes have occurred, because it now can be treated. A drug named D-penicillamine (a breakdown product of penicillin) has the property of binding strongly to copper and promoting its excretion via the kidney, thereby preventing the copper from being deposited in vital organs. In most cases, the effect of penicillamine on the course of Wilson's disease is truly dramatic: many such patients now being followed are leading normal lives, and there are even reports of affected patients having uncomplicated pregnancies and bearing healthy children. Administration of penicillamine must be continued for life, lest copper be permitted to reaccumulate.

Surgery

Surgical repair is most useful in genetic diseases which feature specific nonprogressive physical abnormalities, usually developmental in nature. Thus, a surgical approach is most often employed for the treatment of a number of autosomal dominant and multifactorial conditions which present as "birth defects."

Retinoblastoma is an autosomal dominant disease consisting of malignant tumors in the retinas of the eyes. Formerly, it was either fatal, or was cured only by the removal of both eyes, a most unsatisfactory alternative. However, recently developed surgical techniques make it possible to give a far more cheerful outlook to these patients and their families. Most often, the tumor in one eye is farther advanced than the growth in the second eye; therefore, only the first eye is removed. The smaller tumor in the second eye is then treated by cryosurgery: the application of tiny probes which freeze and kill the tumor cells without affecting the surrounding normal retinal tissue, thereby preserving the child's sight. After surgery, the patient usually is also given a course of anticancer drugs, to wipe out any persisting cancer cells. Sometimes, by good luck, the retinoblastoma is discovered early enough to permit the use of cryosurgery in both eyes. As fine an advance as cryosurgery is, it is not likely to be the last word. In the future, micro-laser beams may provide even greater efficiency in the eradication of retinoblastomas.

Hereditary polyposis of the colon is caused by an autosomal dominant gene mutation. In this condition, the lining of the large intestine is transformed into millions of polypoid tumors, so that microscopically it has the appearance of a shag rug. Unfortunately, the polyps have a high rate of malignant degeneration, so that by the age of forty — and sometimes much earlier — almost every affected patient will have developed intestinal cancer, with its attendant high mortality rate. The polyps can be diagnosed by sigmoidoscopy and/or barium enema, and most geneticists recommend that these procedures be performed every three to five years in persons at risk. When polyps are discovered, thereby establishing the diagnosis, the patients are advised to undergo removal of the large intestine (colectomy). This necessitates lifelong colostomy maintenance, but most people adjust well to this, and do not find it unduly handicapping. It's usually perceived as highly preferable to developing intestinal cancer at an early age.

The majority of single congenital physical abnormalities are caused by multifactorial mechanisms based on polygenic inheritance. Examples are congenital heart defects, and cleft lip and palate ("hare-lip"). Not so long ago, a diagnosis of congenital heart disease was a death sentence. An individual who did not expire as a baby was doomed to a sad, short existence confined to house, bed, and hospital. Now, repair of the majority of these malformations is considered routine surgery. Abnormal communications between heart chambers ("holes in the heart") are easily and permanently closed. Congenital defects of the heart valves or the great vessels leading to and from the heart are more troublesome, but even here, much can be accomplished at this time, and it is likely that the near future will bring improvements. As for cleft lip and palate, many of these babies also used to die, because of inability to take nutrition. Even when repair was attempted, the result usually was poor, leaving the survivor with an unsightly cosmetic problem and unintelligible speech. Present techniques, however, make good use of principles of plastic surgery to produce a repair that is highly satisfactory both functionally and structurally. It should be mentioned in passing that before undertaking surgical repair for cleft lip and palate (and indeed, for all other birth defects), the surgeon must perform a careful examination to make certain that there is in fact only a single defect, and that other anomalies do not coexist. For example, cleft lip and palate is sometimes present as part of the syndrome associated with trisomy for chromosome number 13; the major feature of this condition is terribly severe brain malformation, such that most people would not choose to repair the associated cleft lip and palate.

Surgical repairs usually are of little help in dealing with the biochemical aberrations caused by inborn errors of metabolism. But there is an exception. Glycogen storage diseases are inborn errors that result from the inability to break down glycogen, a compound sugar, into its component glucose molecules. Under normal conditions, glucose is absorbed from the gut, and transported via the blood to the liver, where it is polymerized to glycogen, a convenient form for storage. Later, the stored glycogen can be gradually converted back to glucose, to keep pace with the continual need of the body for this carbohydrate.

Patients with glycogen storage diseases have no problem converting glucose to glycogen. However, the reverse conversion is prevented by deficits of one or another of the numerous enzymes that catalyze the glycogen-to-glucose metabolic pathway. Thus, the net effect is excess storage of glycogen, and deficient release of glucose. Depending upon which specific glycogen-breakdown enzyme is defective, the clinical disease may be more or less serious. Some forms of glycogen storage disease are manifested only by minor enlargement of the liver and periodic mild bouts of low blood sugar (hypoglycemia); these are compatible with normal survival. Other forms, however, are associated with severe liver enlargement, chronic and disabling hypoglycemia, and other, secondary, physical defects. Such patients do not survive childhood.

In a few severely affected children, surgeons have performed interchanges between the hepatic portal vein and the vena cava. This causes blood from the gut, heavily laden with glucose, to bypass the liver and go directly to the heart, from where it is pumped to the rest of the body. Blood to the liver then comes from the legs, and is low in sugar content. This interchange makes more glucose available to the tissues of the body, and less available for storage in the liver. Although the surgery is difficult and dangerous, when it is successful it seems to produce marked improvement in growth, liver size, and general health. The treatment certainly cannot be considered routine, and should not be recommended for the milder forms of glycogen storage diseases, but for the otherwise-fatal severe types, it definitely seems worthwhile.

Exposure to
Idiosyncratically Dangerous Substances

Several gene mutations exist which ordinarily are harmless, but which can be extremely dangerous under certain circumstances, usually exposure of the individual to a specific drug. The study of such conditions is called *pharmacogenetics.* The hazards of giving sulfa drugs to patients with X-linked glucose-6-phosphate-dehydrogenase deficiency were mentioned in the previous chapter. Other examples of similar disorders are porphyria and pseudocholinesterase deficiency.

Acute intermittent porphyria, an autosomal dominant condition, is caused by a defect in the formation of hemoglobin precursors. No specific enzyme deficit has been identified, but the patients are subject to episodes of massive overproduction of porphyrins, which are metabolic intermediaries in the pathway to hemoglobin synthesis. During these episodes, the urine turns dark, and the patients frequently exhibit bizarre, even psychotic behavior. Sometimes, they have such severe abdominal pain that unsuspecting physicians are misled into diagnosing appendicitis, and operating.

Most porphyric individuals enjoy good general health. Their attacks usually are infrequent and mild, unless there is exposure to a "triggering drug." The most common of these medications are the barbiturates. A single dose of phenobarbital can in some unknown way set off an extremely severe attack, and if a patient already in the throes of a porphyric crisis is given phenobarb "to calm him down," it may calm him thoroughly, and permanently. Porphyrics are well advised to wear Medic-Alert bracelets, so that they will never inadvertently be given a dose of barbiturate.

Pseudocholinesterase deficiency is a fascinating condition. Its inheritance is complex, involving several different alleles, but basically, it is an autosomal recessive. Patients deficient in pseudocholinesterase are perfectly healthy, and suffer no disadvantage. There is, however, one exceptional situation, and that is a very important one. To make it easier and safer to give general anesthesia for surgery, a drug named succinylcholine is administered. This agent paralyzes the patient's muscles, including those used in breathing, so that the anesthesiologist can pass the necessary breathing tube into the patient's windpipe. Ordinarily, the effect of succinylcholine is transient, since the drug is rapidly broken down by the pseudocholinesterase in the patient's blood. However, pseudocholinesterase-deficient individuals cannot degrade succinylcholine, and so remain unable to breathe until their kidneys have excreted the medication, usually a matter of several hours. In past years, such patients constituted a proportion of those unfortunate persons who "never woke up from surgery." Today, however, all properly trained anesthesiologists are aware of the existence of pseudocholinesterase deficiency, and when a patient fails to establish proper respiration after surgery, his blood is tested for pseudocholinesterase activity. Meanwhile, the anesthesiologist continues to maintain respiration by machine. If pseudocholinesterase deficiency is indeed the problem, artificial respiration is continued until the succinylcholine has been excreted, and the patient suffers no damage.

We can be certain that the field of pharmacogenetics will continue to enlarge the list of disorders which includes G-6-PD deficiency, porphyria, and pseudocholinesterase deficiency. Already, there are approximately twenty-five known "diseases" of this sort. In addition, most unusually severe drug reactions are probably genetically determined. There are vast chemical differences between individuals, and it is predicted that one day, we will all have our own "metabolic fingerprints." Because this would facilitate early, presymptomatic diagnosis of diseases, such an eventuality should be far more useful to us as individuals than to the F.B.I.

Second-Level Genetic Therapy

These modalities are advanced both a step closer to the actual gene defect, and a step ahead of the observable manifestations of disease. They represent attempts to avert symptomatic illness by the direct administration to the patient of normal gene products designed to do the work of the nonexistent or defective substances coded for by the patient's own genes. Currently, the great majority of available and experimental second-level gene therapies involve the use of enzymes.

By far the best-known example of this type of therapy is the administration of insulin to diabetics. Insulin is a special type of enzyme known as a *hormone:* a compound produced by a particular tissue and then transported by the blood stream to other specific tissues, where it exerts its biological effect. Insulin is produced by the pancreas, and acts at the membranes of the cells of many organs. Its purpose is to regulate certain metabolic processes, especially of sugars, but also of fats and amino acids. Diabetes results from underproduction of insulin, so that sugar (glucose) cannot gain access into the cells of the body; therefore, the blood sugar levels become very high, and sugar "spills over" into the urine. Insulin injections "open the cellular doorway" to glucose, but do not seem to completely prevent other problems characteristic of diabetes, the most prominent of which is increased susceptibility to heart and blood vessel disease. Notwithstanding this shortcoming, insulin has given countless additional years of healthy life to innumerable diabetics since 1922, when the hormone was first used clinically.

A more recent development in enzyme therapy has been the administration to hemophiliacs of Clotting Factor VIII. At least fifteen different factors must interact in order for blood to clot properly; deficiency of any one of them can cause uncoagulability of the blood. Hemophilia — Factor VIII deficiency — is the best known and one of the most common of this group of diseases.

Although it is still an expensive and difficult process, Factor VIII can be separated from banked blood, and then frozen. When administered intravenously to a hemophiliac, it will arrest a bleeding episode. Some hemophiliacs are now being taught to give the compound to themselves two or three times a week; thus, they will always have sufficient circulating Factor VIII to prevent the onset of bleeding episodes. Currently, the major problems associated with this approach are the scarcity and the expense of the Factor VIII preparations.

This discussion of the use of insulin and Factor VIII leads into consideration of one of the major hurdles to be overcome before enzyme therapy can be considered a generally useful modality. Our bodies possess the capacity to recognize as "foreign" any proteins and other large molecules which we ourselves do not produce, and which gain access to our blood or tissues. We react to these molecules by forming appropriately named substances called antibodies, which bind to the "foreign" molecules and aid in their destruction. This cellular xenophobia constitutes an integral part of the body's defense mechanism against invading bacteria and other micro-organisms.

In both diabetes and hemophilia, the enzyme defect is rarely total. Most patients produce a small amount of functional insulin or functional Factor VIII — too small an amount to prevent disease — but enough to prevent the replacement enzyme from being recognized as foreign. However, in the unusual hemophiliac with *total* absence of Factor VIII replacement therapy soon becomes impossible, because of immediate deactivation of the administered Factor VIII by the patient's antibody system.

This same situation probably will be encountered with any inborn error of metabolism where there is absolutely no circulating functional enzyme. In this circumstance, normal enzymes would be recognized as foreign, and destroyed before they could produce their desired biological effects. Hence, mechanisms must be sought to either interfere with specific antibody production or to deliver the enzymes to their target tissues without allowing them to contact antibody-producing cells. Our knowledge of the basic nature of the *immune* (antibody-producing) *system* is so rudimentary that we should not expect to see major breakthroughs in this sphere in the near future.

Another general difficulty relating to the use of enzyme therapy is the fact that normal persons simultaneously produce and inactivate enzymes at similar, relatively rapid rates, so as to maintain circulating levels within the narrow range considered normal. In inborn errors of metabolism, a particular enzyme cannot be manufactured, but there is nothing wrong with the metabolic pathways for enzyme degradation. Therefore, even aside from the question of antibody production, it becomes understandable why hemophiliacs must receive Factor VIII infusions every two or three days to prevent bleeding episodes.

This problem is further illustrated by the experimental intravenous administration of specific enzyme preparations to several patients suffering from Gaucher's and Fabry's diseases, two serious inborn errors of lipid metabolism. In each patient, the enzyme rapidly vanished from the blood, but for a few days afterward, the levels of lipids stored in different tissues showed a hopeful decline. Data such as these underscore the need to develop techniques by which enzymes may be delivered to their target tissues without being exposed to either normal degradative agents or destructive antibodies. Perhaps the enzymes could be sent through the circulation in a nonsoluble form: for example, Dr. K. Adriaenssens of Antwerp has been conducting experiments in which enzymes have been enclosed in the membranes of red blood cells emptied of their normal contents. Until such modifications can be satisfactorily engineered, most enzyme therapies will probably remain at best semiexperimental.

Many enzymes are not simple proteins, but complex substances composed of a protein in combination with a vitamin. In such an arrangement, the protein is called an *apoenzyme* and the vitamin a *coenzyme.* Together, they form a *(holo)enzyme.* Several rare inborn errors of metabolism have been identified as vitamin-dependent inborn errors, since they result from the faulty interaction of a particular protein apoenzyme with its vitamin coenzyme. Sometimes, this defect can be overcome by administration of massive doses of the vitamin, thereby, by sheer force of available numbers, facilitating the interaction of the vitamin with its imperfect apoenzyme. This results in improved enzymatic function and amelioration of disease symptoms.

In no way should this sort of therapy be considered as a rationale for indiscriminate megavitamin dosage of well persons. No scientific evidence exists to support the thought that such behavior will result in better general health; in fact, large doses of vitamins may impair the health of the taker, and, in addition, seem capable of producing birth defects if the vitamin faddist happens to be pregnant.

Surgical techniques have been employed in efforts to provide enzyme therapy. Kidneys have been transplanted into patients with Fabry's disease; this has been done for a dual purpose. First, kidney failure is the usual cause of death in this disorder, because stored lipids accumulate in great quantities in the kidney. Second, the normal kidney is one source of ceramide trihexosidase, the enzyme that is deficient in Fabry's disease. Therefore, the hope has been expressed that implantation of a kidney from a person without Fabry's disease might kill both birds, and save the patient. To date, results have been equivocal: one study appeared to show that the transplanted kidney was producing the enzyme, but a second study came to the opposite conclusion. The doctors and patients in both studies had to deal with the usual problem of rejection of the "foreign" kidney. If we can learn to counteract and control the immune antibody system, surgical transplantation might develop into a major form of enzyme therapy. Normal bone marrow could be transplanted into patients with sickle cell anemia and other diseases caused by abnormal hemoglobins. Pancreatic tissue could be given to patients with diabetes. The list is endless. But unfortunately, right now, the mysteries of immune reactions seem nowhere near solution.

Third-Level Genetic Therapy

Dietary alterations, drug administration or avoidance, surgery, and enzyme therapy are all useful in treating some genetic diseases. Their importance probably will increase with time. But the need for them would virtually disappear if and when it becomes possible to treat genetic diseases directly at the genetic level: by altering the structure or content of cellular DNA. Such therapy would be in the nature of a true cure; furthermore, if it were to include correction of the DNA in the reproductive cells, it would also prevent the transmission of disease-causing genes to subsequent generations.

Direct gene therapy is not a practical therapeutic modality today, nor is it likely to become one tomorrow. Its widespread application most likely will benefit persons we will never know. At this time, the theories related to the future human utilization of third-level treatment modalities are based to a great extent upon evidence gained from genetic experiments which have been performed only on bacteria. Some scientists have expressed the concern that these data may not be transferable to mammals. Keeping this negative thought in mind, we may consider some of the concepts that are likely to lead to the mitigation of a good deal of distress in our distant descendants.

Correction of Defective Genes

When bacteria are exposed to a solution containing DNA extracted from another bacterial species, they will sometimes take up and incorporate the DNA into their own chromosomes (Fig. 4.3), a phenomenon known as *transformation*. The classic experiment proving this fact was done in 1944 by Drs. Avery, MacLeod and McCarty. They exposed a nonvirulent strain of pneumococcus to a solution of DNA extracted from a distinguishable virulent strain. After subsequent cell division, it was discovered that some of the harmless pneumococci had become virulent: they were now capable of causing death when injected into mice. Pretreatment of the DNA extract with DNA-destroying enzymes completely blocked this transformation. Thus, the DNA had succeeded in furnishing to the hitherto nonvirulent bacteria an entirely new (for them) genetic trait: the ability to produce a lethal toxin. It can be seen, then, that the procedure would qualify as a "gene transplant."

Unfortunately for would-be genetic therapists, transformation is an inefficient method of getting new genetic material into a cell; of millions of nonvirulent bacteria exposed to a transforming DNA, only a small number actually take up and incorporate the DNA as part of their chromosomes. For this reason, a good deal of the attention of futuristically inclined geneticists is focused instead on transduction.

Correction of Defective Genes by DNA Transduction

Transduction may be defined as the introduction of new DNA into a bacterial cell by means of a virus; the genetic material becomes permanently incorporated into the cell. Before I discuss this concept, though, it would seem reasonable to first define a virus.

The majority of scientists consider viruses to lie between the living and the inanimate worlds. Most viruses consist of relatively short lengths of DNA — sometimes as short as only three genes — which have somehow managed to surround themselves with protective coat-

TRANSFORMATION OF DNA

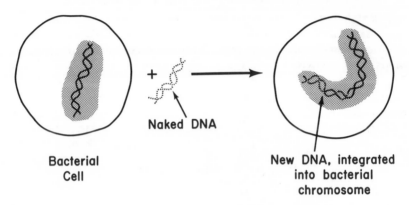

Naked DNA

Bacterial
Cell

New DNA, integrated
into bacterial
chromosome

Fig. 4.3:   Transformation of DNA by bacteria, the process by which fragments
of naked DNA in a solution may be taken up by a bacterial cell and
made part of the bacterial chromosome.

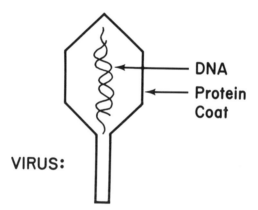

DNA

Protein
Coat

VIRUS:

Fig. 4.4:   The structure of a virus.

ings of protein (Fig. 4.4). Viruses lack the cellular machinery necessary
to metabolism and reproduction, and so must lead a parasitic existence.
Their usual *modus vivendi* (or, strictly speaking, *semi-vivendi*) is to
attach to a living cell and inject their DNA. Some viruses attack
bacterial cells; others attack specific mammalian species. The viral
DNA makes its way to the cell nucleus, attaches itself to a host
chromosome, and uses the metabolic factory of the host cells to direct
both its own replication and the formation of new protein coating.
After assembly of many new viral units of DNA and protein, the
invaded cell bursts, releasing the viruses to go to work on new cells.

How do we get rid of these parasitic monsters before they invade and destroy every cell in our bodies? Sometimes we can't, and then we die, for example as the result of an overwhelming attack of influenza or hepatitis. But usually, we form antibodies to the (foreign) viral protein soon enough to wipe out the viruses before they do us in.

At this point, you're probably wondering what all this has to do with gene transplants. The association is based upon the fact that some viruses behave in a more diplomatic fashion. Rather than mounting an armed invasion and takeover, with a fight to the death, they simply inject their DNA, and ask to live and let live. Some bacterial viruses integrate their DNA into the bacterial chromosome, and others form separate little genetic bundles *(episomes)* outside the nucleus, in the cytoplasm of the bacterial cells. In either event, this type of virus does not disrupt general cellular metabolism, but quietly reproduces its DNA whenever the bacterium reproduces (Fig. 4.5). Such viruses are called *temperate viruses,* and the described phenomenon of coexistence is termed *lysogeny.* Genetically, the result is the same as that of transformation; the cell has acquired new, permanent hereditary material. In man, the herpes ("cold sore") virus is thought to exist in a lysogenic state, the periodic flareups of sores representing short-lived reversions by a few viruses to the destructive condition. These renegades are then neutralized by antibodies, leaving the more reasonable members of the viral population to continue along their peaceful, lysogenic way.

VIRAL TRANSDUCTION OF DNA

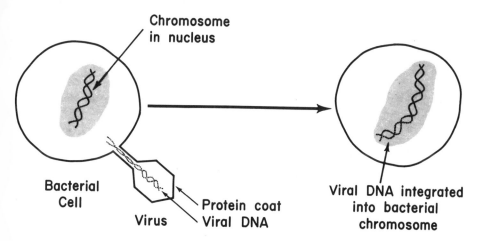

Chromosome in nucleus

Bacterial Cell

Virus

Protein coat

Viral DNA

Viral DNA integrated into bacterial chromosome

Fig. 4.5: Transduction of DNA via a temperate virus.

So: why not find a virus which carries a useful trait, inject it into a person with the corresponding genetic deficiency, and let the virus direct the patient's cells to produce the needed material? A brilliant new idea? Hardly. Since 1798, we've been following the lead of Edward Jenner, by injecting harmless strains of pox viruses into ourselves, so that we may possess antibodies to repel deadly smallpox infections. It is uncertain whether the vaccinating virus is wholly destructive or settles down to a lysogenic existence, but in either case, the fact remains that specific DNA is used to provide us with a new and useful genetic skill.

Can lysogenic gene transduction be applied to "classical" genetic diseases? As far as chromosomal and multifactorial traits are concerned, the answer is probably no. There are too many genes involved, and in most cases, irreparable damage is done early in embryonic development. It is for the inborn errors of metabolism, determined by single gene mutations, that transduction appears to hold the greatest promise. The corrective gene might be transduced during the newborn period, since in most cases, maternal metabolism seems to be protective to the fetus, with damage not occurring until postnatal life. Alternatively, an embryo could undergo its first few days of development outside the mother's body (see chapter 8), during which time the desired gene could be transduced.

The possible utility of gene therapy by transduction is illustrated by experiments related to the diseases galactosemia and hyperargininemia.

Galactosemia is the disorder that results from deficiency of an enzyme which should take part in the conversion of galactose to glucose. Using standard laboratory techniques for cell culture, a team of scientists at the National Institute of Mental Health have grown cells from the skin of a galactosemic patient. After demonstrating that these cells indeed could not metabolize galactose, the researchers infected the cells with a transducing virus known to possess the gene for the enzyme lacking in galactosemia. Follow-up tests indicated that the hitherto defective cells had acquired the capability to handle galactose properly.

Hyperargininemia is a rare inborn error caused by the lack of an enzyme necessary to metabolize the amino acid arginine. The resulting high levels of arginine in the body produce a disease characterized by severe mental retardation, spasticity, convulsions, and death by early adulthood. One possible mode of genetic therapy for this condition involves the use of the Shope papilloma virus (SPV), an agent which causes wartlike skin growths in wild Kansas cottontail rabbits. SPV is known to induce the cells it infects to synthesize an arginase, which is the enzyme lacking in hyperargininemia; hence, infected rabbits have very low levels of arginine in their blood. This fact led to the discovery that many scientists who had worked with SPV also had very little circulating arginine. Therefore, it was supposed that the scientists had been infected with the virus. That this was a harmless situation was suggested by the fact that there could be found no excess illness or death among those workers, some of whom had last been in contact with the virus as long as twenty years before testing.

Proceeding from the above data, injections of SPV were administered to two teen-aged German sisters with far-advanced hyperargininemia, both of whom were deteriorating rapidly. Although some decline in blood arginine levels was noted, the progress of the disease was unchanged. However, further research into the nature of the basic genetic and biochemical defects of hyperargininemia may permit SPV or similar agents to be used with more success in the future.

In the clinical use of transduction, we would not necessarily be limited to nature's supply of remedial genes. As will be described in chapter 11, some small genes have already been synthesized. This opens the door to the possible development of therapeutic "pseudoviruses." The basic unit of such structures would be a virus demonstrated to be both harmless and efficient at integrating itself into human chromosomes. To this virus would be attached any one of a number of made-to-order synthetic genes. The combined "pseudovirus" would then be injected into the deficient embryo or person, hopefully to transduce the patient into good health.

It must be emphasized that procedures of this sort are now in the earliest experimental stages, and long before they can be thought of as routine clinical tools, many questions must be answered. Viruses have been implicated in the causation of some forms of cancer, and it will have to be ascertained that our selective lysogenic friends do not also predispose to malignant cellular changes. We'll want to know whether the integration of a virus into a chromosome has adverse effects on any other aspects of cellular metabolism. We need to discover which mode of transduction therapy is both safest and most efficient: injection of a virus into an embryo; injection into a newborn infant; or exposure to the virus of cultivated cells removed from a baby, with subsequent reintroduction of the "cured" cells into the child's body. To answer these questions will take years of experiments involving bacteria, small laboratory mammals, monkeys, and hopelessly ill, otherwise doomed humans.

During the past couple of years, a group of microbial and biochemical geneticists headed by Dr. Paul Berg of Stanford has been calling for extreme caution with regard to gene transfer experiments that would confer upon bacteria either resistance to antibiotics or the potential to cause cancer. Scientists who work in these areas met in California in February, 1975, at which time they put together a list of restrictions and guidelines to maximize potential benefits and keep to a minimum the possibility of developing a real-life Andromeda strain. It should be emphasized that Dr. Berg's group has not recommended

EXPERIMENTAL CHIMERA FORMATION: CELL INJECTION

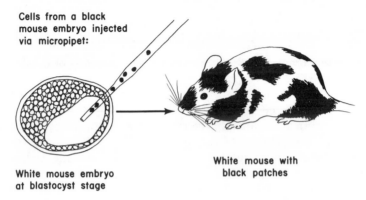

Cells from a black
mouse embryo injected
via micropipet:

White mouse embryo
at blastocyst stage

White mouse with
black patches

Fig. 4.6a: Production of a chimera by injection of cells from one embryo into another, at the blastocyst stage.

EXPERIMENTAL CHIMERA FORMATION: CELL FUSION

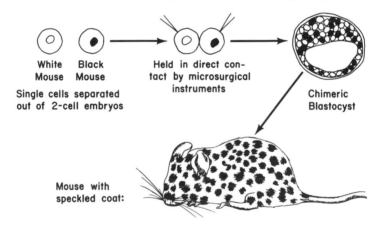

White   Black
Mouse   Mouse

Single cells separated
out of 2-cell embryos

Held in direct con-
tact by microsurgical
instruments

Chimeric
Blastocyst

Mouse with
speckled coat:

Fig. 4.6b: Production of a chimera by direct embryo fusion.

restraints on all transduction experiments, and that experiments falling within the two "high risk" categories comprise only a small minority of the investigations now in process that could lead to direct gene therapy.

Other Possible Modes of Direct Gene Therapy

Groups of genes are like most societies: there are workers, and there are bosses. Some genes turn out the enzymes, and other genes regulate the production rate. Therefore, it has been proposed that inborn errors characterized by total enzyme deficiencies may be due to malfunction of the "boss" genes which should be "turning on" the "worker" genes. Thus, for example, there may be two types of hemophilia: the first type would be caused by an abnormal "worker" gene, which produces an abnormal enzyme with perhaps 3-5 percent activity; the second type would result from an abnormal "boss" gene, so that the "worker" is never told to produce, and there is no enzyme at all. To treat this latter problem, it would be necessary to "switch on" the normal "worker," a process known as *gene de-repression*. How this might be therapeutically accomplished cannot be determined until we acquire a very large quantity of basic knowledge concerning the mechanisms of gene function.

Chimeras are individuals made up of cells from two separate fertilized eggs. As shown in Figs. 4.6a and 4.6b, chimeras can be constructed experimentally either by injection of cells from one embryo into another, or by direct fusion of two embryos. In either case, the result is a healthy, fertile animal whose body contains freely intermingled cells from both original embryos. Unfortunately, chimera production by cell injection is impossible after the patient has been born, since he will then reject the "foreign" cells.

However, embryos do not recognize cellular material from other embryos as "foreign," and so will accept the grafts, and not form antibodies to them. It has been suggested that such a procedure might be of therapeutic use, in that cells from a genetically healthy embryo could be combined with cells from an embryo which would otherwise develop a genetic disease. On the cellular level, the resulting individual would be somewhat akin to a female carrier of an X-linked recessive disorder.

Other procedures have been mentioned as having potential for genetic therapy, such as alteration of individual genes by laser beams, and techniques for manipulating individual chromosomes. However, right now, such concepts are barely beyond the status of science fiction.

The Present Major Accomplishment of Gene Therapy

Perhaps it would be appropriate to conclude this chapter by saying a few words about the modern therapy of Rh-disease of the fetus. Like the A-B-O factors, the Rh factor is a protein located on red blood cells. It is present on the red cells of about 85 percent of humans. Possession of the Rh factor is an autosomal dominant trait, so that persons with either one or two genes for the production of Rh protein are Rh positive. Persons with two "nonproducing" genes are Rh negative, and if Rh positive cells enter their circulations, they will consider these cells as "foreign," and form antibodies to destroy them.

Thus, when an Rh negative woman conceives by an Rh positive man, the stage is set for trouble. If the fetus inherits an Rh positive gene from its father, and if a quantity of its blood enters the mother's circulation (which occurs most frequently at delivery, when the placenta is torn free), the mother will become *sensitized*, that is, she will produce antibodies to destroy the invading red cells. Of course, this does not harm the baby, who by then already has been born. However, if the mother should again become pregnant with a second Rh positive fetus, her previously alerted immune system will pour out anti-Rh antibodies, which then will cross into the fetal circulation and destroy the fetus' red cells. This will cause the fetus to become anemic, often sufficiently so that it succumbs.

Modern therapy for this condition uses the "fight fire with fire" approach. To prevent the initial disastrous maternal sensitization, an Rh negative mother is given an injection of anti-Rh antibodies immediately after the birth of an Rh-positive baby. These injected antibodies destroy any Rh positive fetal cells in the mother's circulation *before* the fetal cells can stimulate the mother to form her own antibodies. Therefore, in the next pregnancy, the fetus will not be exposed to the continuous production of maternal antibodies. The injected antibodies pass out of the mother's body within a few months, and so do not pose a threat to future pregnancies. The injection must be repeated after each pregnancy at risk.

How valuable is this therapy? Formerly, about one baby in 100 was affected to some extent by Rh-disease. Many of these infants were stillborn or died shortly after delivery. Some of the survivors suffered brain damage. The injections appear to be better than 99 percent effective in preventing maternal sensitization. (The occasional failures are probably due to the entry into the maternal circulation of quantities of fetal blood so large that complete neutralization by injected antibodies is impossible, or to crossing over of fetal blood earlier in pregnancy, due to various abnormal pregnancy conditions.) Thus, universal therapy by antibody injections could lower the incidence of Rh-disease of the fetus to under one case in 10,000 newborns, which would make this genetic scourge less common than PKU. That would be a major accomplishment indeed.

5

The Prenatal Diagnosis
of Genetic Disease

Until the late 1960s, genetic counseling was usually a frustrating experience. Couples at risk could be told the name of the genetic disorder that might appear in their offspring, along with the clinical manifestations of the disease. In most cases, the counselor could also make an accurate estimation of the magnitude of the risk.

Faced with these facts, such couples had only two options. Based on the clinical severity of the disease, the magnitude of the risk, and how badly they wanted to have children, they could either go ahead and hope for a lucky roll of the genetic dice, or they could decide to remain childless. All together, this was an unsatisfactory situation.

But changes were in the making. Ten years ago, efforts to liberalize abortion laws were getting under way, and within a few years, most people had come to agree that abortion should be permitted where there existed a good chance that the child would suffer serious mental or physical impairment. Then, in 1966, two groups of scientists simultaneously reported that it was possible to perform chromosomal analyses of fetal cells removed from the amniotic sac. The next year, both Dr. Carlo Valenti of Brooklyn and Dr. Henry Nadler of Chicago used this new technique to diagnose Down's syndrome (mongolism) in a fetus. The two pregnancies were terminated, and in each case, the diagnosis of Down's syndrome was confirmed by examination of the abortus.

To date, this has been the greatest breakthrough in the field of medical genetics. At last, genetic counseling possesses a practical technique to permit at least some couples at risk to have children free of the fear that the offspring might be gravely malformed or mentally retarded.

Historical Development of Prenatal Diagnosis

The prenatal diagnosis of genetic characteristics has been of interest to man since antiquity. Most of the early attention focused upon attempts to diagnose sex. Ancient Egyptians believed that the urine of a woman carrying a male fetus would germinate wheat seeds, while urine of a woman with a female fetus would germinate barley. Hippocrates associated male pregnancies with good color of the mother and with swelling of the right breast and nipple; female pregnancies were supposed to cause skin pallor and swelling of the left mammary apparatus. In the second century A.D., Soranus stated that male fetuses were associated with strong *in utero* movements and that women carrying girls suffered from excessive vomiting.

Many such beliefs persist today. I've seen several patients who have insisted that the fetal sex can be accurately predicted by the loudness of the baby's heartbeat or by the rotation of a needle suspended from a thread and held over the pregnant uterus.

Innumerable superstitious practices have long been employed in attempts to ward off birth defects, but until recently, the antenatal diagnosis of such problems has been a concept far beyond available methodology. In 1900, the British obstetrician Ballantyne could speak of attempts to diagnose both fetal heart disease and anencephaly by the character of the heart sounds. This represented a tentative beginning.

Prenatal diagnosis turned the corner in the late 1950s and early 1960s, with the work of Dr. D. C. A. Bevis in Britain, and Dr. A. W. Liley in New Zealand. These physicians performed *amniocentesis* (transabdominal puncture of the amniotic sac) to obtain fluid which they then analyzed for the quantity of degraded blood pigments. These data allowed them to determine the severity of Rh-incompatibility disease of the fetus. A few years later, Valenti and Nadler combined the technique of amniocentesis with recent major advances in laboratory cell culture and chromosomal analysis to make the first prenatal diagnosis of genetic disease.

Today, amniocentesis remains the primary tool in prenatal diagnosis. Other currently utilized techniques are ultrasound scanning and X-ray.

Prenatal Diagnosis Techniques: Amniocentesis

Amniocentesis should not be considered an entity unto itself, but a technique used in prenatal diagnosis, which is itself a branch of genetic counseling. Amniocentesis or other prenatal diagnostic procedures should never be performed without antecedent thorough genetic counseling by someone expert in the field. Most patients are not familiar with the risks and problems of amniocentesis itself. Some patients do not realize that abortion is the only alternative should chromosomal or most biochemical diseases be diagnosed; they believe that intrauterine therapy is possible. Genetic counseling permits the couple at risk to ask questions, consider the alternatives, weigh the risks, and finally come to an informed decision. Only then should investigative procedures be undertaken (Fig. 5.1).

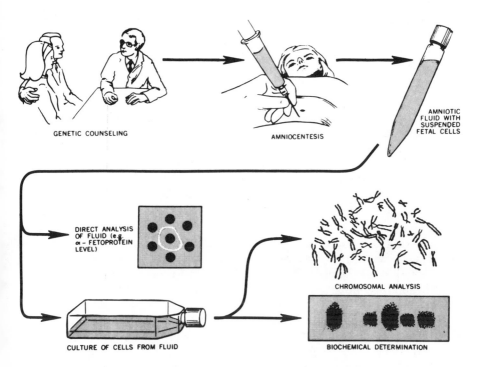

GENETIC COUNSELING

AMNIOCENTESIS

AMNIOTIC FLUID WITH SUSPENDED FETAL CELLS

DIRECT ANALYSIS OF FLUID (e.g. α – FETOPROTEIN LEVEL)

CHROMOSOMAL ANALYSIS

CULTURE OF CELLS FROM FLUID

BIOCHEMICAL DETERMINATION

Fig. 5.1: The proper sequence of events in prenatal diagnosis. Genetic counseling should always precede the performance of any diagnostic maneuvers. **(Courtesy Randy Kobernick, editor,** *University of Washington Medicine.)*

Amniocentesis was long considered dangerous, and was used only sparingly and as a last resort, usually in hydramnios, a condition of obscure causation in which dangerous excesses of amniotic fluid accumulate. But when Liley and Bevis popularized the use of amniocentesis in the assessment of Rh-incompatibility disease, it was realized that the procedure was relatively innocuous. So, the use of amniocentesis was extended to include assessment of fetal maturity, by analysis for creatinine content and for the presence of mature skin cells.

Clinical and Laboratory Techniques of Amniocentesis

For the assessment of both fetal maturity and the severity of Rh-incompatibility disease, amniocentesis is performed during the later weeks of pregnancy. However, to detect genetic disease in the fetus, the tap must be done between fourteen and twenty weeks after the last menstrual period. At this time, there is enough amniotic fluid to permit easy removal of a sample, but it is still early enough to perform abortion, if disease is discovered.

After appropriate genetic counseling, the patient is instructed to empty her bladder. This is done so that the bladder will not obstruct the path of the needle to the uterus, and also because amniotic fluid at this stage of pregnancy looks exactly like clear, yellow urine. Next, a pelvic examination is performed, to determine the size and position of the uterus, and the optimal site for puncture. During the examination, if an area of unusual softness and prominence is noted in the uterine wall, this region is avoided, since it probably represents the placental implantation site. Many doctors routinely localize the placenta by ultrasound, but the necessity of doing this has not been established.

The patient's lower abdomen is then cleaned with an antiseptic solution, and the skin and the underlying lower abdominal wall are injected with a local anesthetic. After a few minutes, a four-inch-long needle is passed through the anesthetized region and into the uterus; as the amniotic sac is entered, the operator usually feels a distinct lessening of resistance or "give" (Fig. 5.2). About twenty cubic centimeters, or two-thirds of a fluid ounce, are drawn into a syringe, and then the needle is removed. The patient sometimes is kept under observation for about an hour after the procedure; if at that time there is no pain, bleeding, or vaginal leakage of fluid, she is sent home to await results.

Amniotic fluid contains small numbers of living body cells. The exact sites of origin of these cells have not yet been determined, but they do arise from the fetus and are probably sloughed off from the skin, the amniotic membrane, the urinary tract, and the respiratory system. When the fluid with its suspended cells is mixed with nutrient-containing tissue-culture fluid, and then maintained in an incubator at 37° C (normal human body temperature), the living cells will attach to the walls of the plastic culture dish and reproduce, forming colonies. Fetal cells grow slowly at first, reaching large numbers in one and one-half to two weeks. Occasionally, the cultures are contaminated by maternal uterine cells, but fortunately, these proliferate more rapidly than fetal cells and then die out in about a week. The cultured fetal cells can be analyzed chromosomally after about two weeks of growth; about four weeks are necessary before enough cells accumulate to permit the performance of tests for biochemical abnormalities.

Although most prenatal diagnoses are made by analysis of the cultured cells, occasionally the fluid itself is utilized.

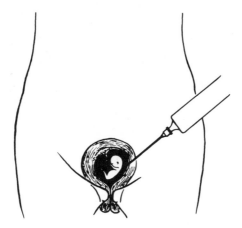

Fig. 5.2:   Technique of amniocentesis. The needle penetrates the abdominal wall and the uterine wall, to enter the fluid-filled amniotic sac.

Problems and Hazards of Amniocentesis

Although it sounds highly distasteful and probably a bit scary to have a long needle thrust deep into one's lower belly, the vast majority of patients have expressed surprise at the lack of discomfort. Most of them have said that it was no more unpleasant than a blood test.

Nor is amniocentesis at fourteen to twenty weeks of pregnancy a particularly dangerous procedure. At first, it was feared that needling the uterus this early in pregnancy might provoke miscarriage. However, it is now apparent that the rate of miscarriage after genetic amniocentesis is no higher than the 1 to 3 percent spontaneous abortion rate at this time of pregnancy.

A more real risk is that of striking the fetus with the needle. This has occurred on several occasions, leaving a small punctate or linear depression in the skin. However, since only the trunk or the extremities have been hit, the scars have been of no importance. Only if the face or an eye were punctured might such an accident assume significance. This has not yet been documented to have occurred in a genetic amniocentesis, and is not likely to happen, since at this point in pregnancy, the fetal body is well flexed, with the face "looking down" at the chest.

Other complications have been reported infrequently. Hemorrhage has resulted from the inadvertent puncture of a large vessel in the uterus or in the abdominal wall, and intrauterine infection has occurred, presumably secondary to bacteria introduced via the needle. In addition, there is concern that Rh-incompatibility disease might be initiated by puncture of a placental vessel, allowing the escape of blood cells from an Rh-positive fetus into the circulation of an RH-negative mother. So far, however, this has not been reported.

Lastly, there is as yet no definite proof that the removal of fluid at this stage of pregnancy is totally without effect on the fetus. Although infants born after amniocentesis seem to develop normally during the first couple of years of life, the concern has been expressed that the sudden change in intrauterine pressure caused by the fluid withdrawal might adversely affect the developing brain, perhaps causing the loss of a few points of IQ. Studies are now under way in many centers to evaluate this possibility. I believe that these studies will provide reassuring information.

Aside from actual and potential risks, there are also a few technical problems associated with amniocentesis.

In less than 5 percent of attempts, the amniotic sac cannot be located with the needle. When this happens, the patient must be seen again a week or two later. By then, the uterus will have increased in size.

In about one percent or less of cases, fetal cells fail to grow out in the laboratory incubator. This may reflect either inadequate numbers of viable cells in the original sample or inadvertent contamination of the fluid with microorganisms. In any event, when it happens, another tap must be done.

If the patient is carrying a twin pregnancy, only one diagnosis will be made, since twins are almost always contained in separate sacs. Therefore, when the likelihood of twins is elevated, or the risk of disease is high, ultrasound scanning of the uterus should be performed to identify the number of fetuses present.

Patients must always be told that there is no prenatal diagnostic procedure, including amniocentesis, that will screen for *all* birth defects. About 3 percent of newborns suffer from major physical or mental abnormalities, most of which cannot be identified before birth. Therefore, the counseled couple must understand that prenatal diagnosis will not guarantee them a normal baby.

In summary, if we add up all the actual and theoretical risks of amniocentesis, we arrive at a total risk figure of approximately 0.5 percent. Admittedly, this is a "soft figure," since, so far, only a few thousand taps have been performed and then reported in the medical literature. But the estimation of risk magnitude is important, since it allows us to conclude that at the present, the performance of diagnostic amniocentesis is certainly reasonable when the risk of diagnosable disease is one percent or higher. As will be seen shortly, this is a significant consideration.

Indications for Amniocentesis

Amniocentesis, the principal technique in prenatal diagnosis, is used to diagnose a wide variety of diseases. These may be grouped into chromosomal abnormalities, X-linked disorders, inborn errors of metabolism (mostly autosomal recessive diseases), and anencephaly/ spina bifida. For the first three groups, diagnosis depends on the testing of amniotic-fluid cells; for the last, the fluid itself is analyzed.

Chromosomal abnormalities. Since chromosomal analysis can be performed on cells recovered from amniotic fluid, any disease caused by numerical or structural abnormality of the chromosomes can be detected prenatally. In fact, the exclusion of Down's syndrome (mongolism) is by far the most frequent indication for amniocentesis. Down's syndrome is caused by a third chromosome number 21 (see chapter 1). Patients at risk for children with Down's syndrome fall into three categories: older mothers, women who have previously borne a child with that disease, and women having two or more close relatives with mongolism.

By "older mothers" is usually meant women at least in their late thirties. For unknown reasons, the incidence of Down's syndrome increases markedly with maternal age. Advanced age of the father does not seem to play a role; this may be due to the fact that sperm are produced continuously during a man's reproductive life, while all the eggs a woman will ever have are present before her own birth. Therefore, it is possible that as the egg ages in the ovary, it becomes defective in such a fashion as to predispose to the inclusion of an extra chromosome near the time of ovulation or fertilization. The striking nature of the maternal age effect is illustrated by the fact that the incidence of Down's syndrome in the offspring of women under thirty is about one in 1,500. This figure rises to one in 300 between thirty-five and thirty-nine; one in 100 between forty and forty-four; and one in forty between forty-five and forty-nine. Some recent data suggest that the risk may be even higher. Thus, the risk for Down's syndrome in the offspring of women in their late thirties and older exceeds the intrinsic risks of amniocentesis. Around the age of thirty-five, the two sets of risks are probably about equal.

Down's syndrome is not the only condition that increases with maternal age; the same relationship holds for the other diseases caused by extra autosomes or extra sex chromosomes that were discussed in the Introduction. These diseases are less common than mongolism, but the clinical pictures of most of them include mental retardation and physical abnormalities. The chromosome analyses performed to rule out Down's syndrome will also detect these diseases, so this increases the advisability of amniocentesis for older mothers.

Once a woman has given birth to a child with Down's syndrome, she is at increased risk for recurrence in subsequent pregnancies. Presumably, she is more susceptible to the unknown factors that predispose to inclusion of an extra chromosome in the egg. It was formerly taught that the risk of recurrence of Down's syndrome was three times the age-specific risk, which would still not be a very high figure for women under thirty. However, results from amniocenteses performed on women with previous Down's babies show a recurrence rate of about one percent. Since all the series included a large proportion of younger mothers, it must be concluded that regardless of maternal age, the previous birth of a child with Down's syndrome constitutes a valid indication for amniocentesis. This philosophy is strengthened by the reassurance factor. In the past, many women with mongoloid children took every effort not to conceive again, and if they accidentally did become pregnant, they resorted to abortion. In my own experience, prenatal diagnosis patients referred because of previous mongoloid children often react to being given a recurrence of "only" about one percent by saying, "I don't care what the numbers are. I won't take *any* chance that it might happen again."

Smaller numbers of patients present for advice because multiple family members suffer from Down's syndrome. A minority of these situations represent familial, inherited translocations, involving attachment of the extra number 21 chromosome to another chromosome. In this case, the risk for a mongoloid child depends upon whether the person seeking counseling is one of the unfortunate family members who happen to carry the translocation in the asymptomatic (balanced) form. However, even when multiple familial cases of Down's syndrome are found to be associated with the usual, free extra number 21 chromosome, probably the consulting patient is still at increased risk. In such a situation, there may exist a familial predisposition to trisomy because of an ill-defined inherited increased likelihood of inclusion of an extra chromosome in an egg. Although no firm risk figure can be given in such a situation, most geneticists believe that amniocentesis is justified where, for example, a sibling and a maternal aunt or cousin both suffer from trisomy-21.

Thus, amniocentesis can be used to diagnose numerical chromosomal abnormalities, most of which are associated with advanced maternal age, and are more likely to appear a second time where they have already occurred once. In addition, amniocentesis offers the opportunity to diagnose structural chromosomal abnormalities, and therefore to recognize the fetus missing part of a chromosome or having duplication of part of a chromosome. Most of these situations are caused by unbalanced translocations, where one or the other parent is an asymptomatic, balanced carrier. The unbalanced offspring usually have serious mental and physical abnormalities. Some unbalanced translocations produce such severe genetic derangements in the embryo that miscarriage results; therefore, the risk for live-born, abnormal babies to balanced translocation carriers varies with the chromosomes involved in the translocation and usually is between one and 15 percent. Therefore, if either prospective parent is known to carry a balanced translocation, the woman should have her pregnancy monitored by amniocentesis.

X-linked disorders. Chromosomal analysis of amniotic cells is also useful when dealing with X-chromosome linked disorders. By far the most common diseases in this group are hemophilia and Duchenne's muscular dystrophy. Both are chronic ailments, requiring frequent and expensive treatment. In addition, muscular dystrophy invariably follows a long, downhill course terminated by bedridden existence, with death at about the age of twenty. Unfortunately, at this time, neither disease can be detected prenatally.

However, the problem can be dealt with by sex determination. Since half the sons of a carrier woman will be affected, chromosomal analysis can be performed to reveal the fetal sex. Using this information, the patient could opt for abortion of all male fetuses, therefore deciding to have an all-girl family. This technique can also be used when carrier status cannot be determined with certainty. For example, when pedigree analysis reveals a carrier likelihood of 50 percent, the risk that a male child will be affected is 25 percent, still a high figure.

A situation involving the abortion of fetuses with at least an even chance of being normal may seem to be a less than satisfactory arrangement, and indeed it is. However, at the present time, it's the best that can be offered for prevention of hemophilia and muscular dystrophy. The large majority of carrier women have closely followed the progress of the disease in a brother or a cousin, and they are certain that they do not want to take a 50 percent, a 25 percent, or even a 12½ percent chance that their own sons will suffer from the disease.

Inborn errors of metabolism. Amniocentesis also is of use in the diagnosis of some inborn errors of metabolism. Most of these are inherited in an autosomal recessive fashion, but a few are X-linked. Inborn errors of metabolism are caused by defective single genes, and thus are not associated with any observable chromosomal abnormality. Diagnosis depends upon the ability to assay the amniotic cells for the specific crucial enzyme whose absence causes the disease, or for the presence of an abnormal compound in the cells which accumulates because the enzyme is not there to metabolize it.

To permit the diagnosis of an inborn error of metabolism, the function of the enzyme in question must be discernible in amniotic cells. For this reason, phenylketonuria cannot be diagnosed *in utero:* the enzyme phenylalanine hydroxylase, whose absence causes the disease, is present only in liver cells. Similarly, the abnormalities of sickle cell anemia are apparent only in the red blood cells of the fetus, so diagnosis of this disease must await the development of a safe and reliable technique to obtain fetal blood samples. Cystic fibrosis cannot be diagnosed prenatally because the responsible enzyme has not been definitely identified, and no reliable test is available for its detection. In addition, no abnormal biochemical compounds are seen consistently in the amniotic cells of fetuses later shown to have cystic fibrosis.

Several metabolic diseases have been diagnosed in the second trimester of pregnancy; in addition, there are perhaps twenty more for which usable assays have been designed (Table 5-1).

Table 5-1

Inborn Errors of Metabolism That Have Been
Diagnosed in the Second Trimester of Pregnancy

Disease	Technique
Disorders of lipid metabolism	*Enzyme Assay on*
Tay-Sachs disease	*cultured amniotic*
Sandhoff disease	*fluid cells*
Niemann-Pick disease	
Metachromatic leukodystrophy	
Gaucher's disease	
Generalized gangliosidosis	
Krabbe's disease	
*Fabry's disease	
Disorders of amino acid metabolism	
Maple syrup urine disease	*Enzyme Assay on*
Propionic acidemia	*cultured amniotic*
Methylmalonic aciduria	*fluid cells*
Argininosuccinic aciduria	
Cystinosis	*Increased nonprotein cystine in*
	cultured amniotic fluid cells
Disorders of carbohydrate metabolism	*Enzyme Assay on*
Glycogen storage disease, Type II (Pompe's)	*cultured amniotic*
Galactosemia	*fluid cells*
Disorders of Mucopolysaccharide metabolism	*Abnormal incorporation of sulfate*
Hurler's syndrome	*by cultured amniotic fluid cells.*
*Hunter's syndrome	*Abnormal levels of sulfated com-*
	pounds in amniotic fluid
Miscellaneous disorders	*Enzyme Assay on*
Lysosomal acid phosphatase deficiency	*cultured amniotic*
*Lesch-Nyhan syndrome	*fluid cells*
Xeroderma pigmentosum	

* X-linked disease. All others are autosomal recessive.

Anencephaly and spina bifida. The amniotic fluid itself is of primary diagnostic value in one situation: the prenatal diagnosis of anencephaly and spina bifida. These are two clinical variations of the same pathological process. The brain and the spinal cord are a continuous structure; together they form the central nervous system. In the early embryo, these structures originate as an elongated groove, which develops into a closed tube during the second month after fertilization. After the formation of this neural tube, the bony skull and vertebral column develop to cover and protect the nervous system. In about one live birth in 500 in the United States, the primitive neural groove fails to close into a tube along its entire length, and therefore is not completely covered by bone. If failure to tubulate occurs at the level of the future brain, further development here remains rudimentary, and the skull does not form. This is called anencephaly (Fig. 5.3). It is invariably fatal before or shortly after birth. If, however, the defect is lower, at the spinal level, it is called spina bifida.

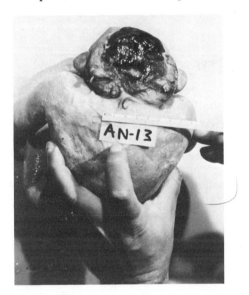

Fig. 5.3:   Anencephaly, back view. Both the disorganized mass of brain tissue and the failure of the skull bones to develop result from a critical developmental error during the second month of pregnancy. (Courtesy Dr. Ron Lemire, University of Washington School of Medicine.)

Many newborns with spina bifida survive. Depending on the size and level of the spinal defect, there will be weakness or paralysis of the legs, inability to gain control of the bladder and bowel, and troublesome spinal curvatures. Because of the opening into the spine and the operations necessary to correct it, meningitis is a frequent complication, In some cases, the defect blocks the flow of cerebrospinal fluid, causing hydrocephalus, or "water on the brain." Some degree of mental retardation is usually present in patients with spina bifida. All told, these patients may exact the highest cost of any birth defect, in both emotional and financial terms.

Anencephaly/spina bifida is inherited by a polygenic mechanism. After the birth of one such baby, the recurrence risk is 5 percent. Half of the risk is for anencephaly, half for spina bifida.

To diagnose these conditions prenatally, the cell-free amniotic fluid is assayed for alpha-fetoprotein, a substance produced by fetal tissues, but whose biological role is unknown. The compound is normally found in small quantities in amniotic fluid. In 1972, the Englishmen Brock and Sutcliffe reported that in the amniotic fluid of an anencephalic pregnancy, alpha-fetoprotein was increased ten times over the normal level. Follow-up work has amply confirmed this observation, and assay for alpha-fetoprotein is now the standard test for anencephaly/spina bifida.

Prenatal Diagnosis Techniques: Ultrasound

Amniocentesis is not the only technique used for prenatal diagnosis. Ultrasound also plays an important role, both as an adjunct to amniocentesis and as a primary procedure.

Medical diagnostic ultrasound operates on the same physical principles as does sonar. A transmitting crystal directs an intermittent ultrasound wave into the patient, and a receiving crystal detects the alteration in wave characteristics caused by an interface of altered density. Such interfaces exist between solid tissues (bone), soft tissues (brain, liver, uterus), liquid (amniotic fluid), and air (abdominal body cavity). The transmitting and receiving crystals are moved in a systematic fashion, horizontally and vertically, over the entire abdominal surface overlying the uterus. This procedure is called scanning. The sound impulses emanating from interfaces are converted to electrical impulses, which in turn are converted to light, and displayed on an oscilloscope screen. In this way, composite outlines are obtained of the uterus and its contents at several levels. Polaroid pictures can be taken for a permanent record.

Some prenatal diagnosis centers use ultrasound routinely to try to localize the placenta before amniocentesis, thereby hoping to avoid placental puncture, which in theory could lead to both miscarriage and to sensitization of an Rh-negative woman to the Rh factor. However, I believe it is yet to be proved that placental puncture is in fact dangerous, and that prior ultrasound identification of the placental site will indeed lessen the incidence of placental puncture. Moreover, although the low levels of ultrasound used for diagnostic purposes do appear to be quite safe, I think it is too early to be certain that the technique is totally without risk. It might be well to recall that in the early days of diagnostic X-rays, it was considered that low levels of radiation were benign. Only years later did it become apparent that the only totally safe level of radiation was zero. Therefore, for the present, I would prefer to use ultrasound only on specific indication: where the patient is RH-negative and therefore definitely susceptible to RH-incompatibility disease, or where the presence of a twin pregnancy would present a problem. An example of this situation would be a pregnancy at risk for hemophilia or muscular dystrophy. Since it would not be possible to tap both sacs, and therefore to be certain both fetuses are female, some patients in this situation might prefer to resort to abortion and then to try another pregnancy.

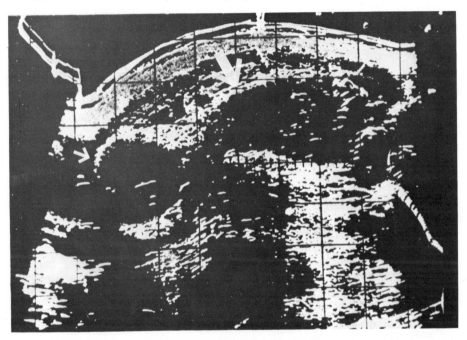

Fig. 5.4: Ultrasonogram showing fetus *in utero*. The round structure (short, thin arrow) with a midline echo is the head; the oblong structure (long, thick arrow) is the trunk. (Courtesy Dr. Sue Conrad, University of Washington School of Medicine.)

Ultrasound can help in the prenatal diagnosis of anencephaly. Since the skull bones fail to form, the characteristic round outline of the skull is absent (Fig. 5.4). Present ultrasound techniques are not sufficiently sensitive, however, to recognize spina bifida early enough in pregnancy to permit abortion.

Ultrasound has also been employed in an attempt to diagnose congenital microcephaly, a condition whose major features are small head size and profound mental retardation. Research performed on fetuses has provided evidence to suggest that in congenital microcephaly, the size of the head may already be small at eighteen weeks of pregnancy, at which time it could be recognized by ultrasound.

Prenatal Diagnosis Techniques: X-ray

X-rays are occasionally called upon in prenatal diagnosis, where the disease in question involves the bones. One such disease is the thrombocytopenia-absent radius syndrome, an autosomal recessive condition. This is a serious illness: the thrombocytopenia (too few blood platelets) results in poor clotting, causing episodes of internal and external bleeding, while the absence of the radius (the bone of the forearm) produces a "seal-like" deformity in which the hand arises directly from the elbow. One of our patients was the mother of an affected three-year-old child, mentally retarded due to a cerebral hemorrhage. She was unwilling to take the 25 percent chance of recurrence, and so decided to have an abortion when she found herself accidentally pregnant. However, at eighteen weeks of pregnancy, we were able to demonstrate the presence of a fetal radius by X-ray. Since bilateral absence of the radius is an invariable feature of the disease, we were able to assure her that her fetus did not have the T-A-R syndrome. She decided to continue the pregnancy and delivered a healthy baby at term.

Controversial Issues in Prenatal Diagnosis

There are many debatable points regarding prenatal diagnosis, and we can be certain that the future will bring more. These questions are difficult to resolve, since they involve moral, ethical, and philosophical viewpoints which are subject to the greatest individual variation. Therefore, I believe that it is the responsibility of the genetic counselor to adopt a neutral attitude as he provides his patients with the factual information they need to make the proper decisions according to their own consciences.

Perhaps the most controversial aspect of prenatal diagnosis is that it leads to the abortion of fetuses doomed to be born malformed and/or mentally retarded. Reams of paper have been filled with thoughtful analyses of the questions as to whether abortion is murder, when the fetus acquires legal rights, and whether women should be forced to continue unwanted pregnancies. The disagreement arises basically from the unresolved issue of when a fetus becomes a human being (whatever that is). The fields of law, religion, and ethics have not been able to provide anything resembling an unequivocal answer: however, many people seem to feel reasonably comfortable using the average time of viability (capacity for independent existence) as the milestone for this fetal achievement. In the face of such widespread, fundamental disagreement, I can only conclude that the issue of abortion ought to remain a matter for individual decision.

A generally unappreciated sidelight to the abortion issue is the fact that prenatal diagnosis undoubtedly saves more fetal lives than it terminates. My own experience and that of other workers in the field is that a large number of pregnant women carrying high genetic-risk pregnancies would resort to abortion if diagnostic techniques were unavailable. This is in contrast to the small number of patients who are discovered to be carrying a seriously defective fetus.

A related question is whether abortion should be performed upon diagnosis of a disease that is treatable postnatally. The critical issues to be considered in this situation are the cost and complexity of the therapy, and to what degree it will permit the patient to lead a normal life. This problem is illustrated by galactosemia, which, untreated, leads to cataracts in the eye, liver damage, mental retardation, and eventually death. However, if milk and milk products are avoided, for example, by the use of soy bean formula in infants, patients with galactosemia can lead quite normal lives. Prenatal diagnosis is possible for galactosemia, and it could be followed either by abortion of the affected fetus or by continuation of the pregnancy and immediate postnatal treatment. Since simple dietary therapy permits galactosemics to live with little or no difficulty related to their disease, most prospective parents decide to continue the pregnancy. On the contrary, the treatment of hemophilia involves frequent, expensive intravenous infusions, and most families choose not to risk this possibility.

Sometimes there is a problem with the so-called unexpected result. A case has been reported involving a forty-two-year-old woman whose fetus was found to have not Down's syndrome, but a 47, XYY, chromosomal arrangement. This is the so-called "supermale karyotype" (see Introduction) first discovered among maximum security prisoners and therefore linked to aggressive, antisocial behavior. Later work has revealed that about one of every 800 men has an extra Y chromosome, and that most of them are perfectly normal, law-abiding citizens. The patient's doctors debated whether to tell her simply that her child did not have Down's syndrome, or whether she should be given all the facts. The doctors believed that an XYY karyotype did not constitute a good reason for abortion, and that if the patient were to continue the pregnancy, her relationship with her child might be severely affected in that she might interpret as pathological every normal act of childhood aggression. In the end, it was decided that not to tell the patient everything would constitute unethical withholding of information. I think that this was the proper decision. The patient should be entitled to all information obtained, and it ought to be the continuing responsibility of her physicians to help her understand it and cope with it. Furthermore, the patient, not the doctor, must have the final say as to which condition would justify abortion *for her*. Other sex chromosomal abnormalities, such as 47, XXX, are associated with relatively low rates of mental and physical problems. Most patients would want to continue such pregnancies, but others might not wish to risk any deviation from a normal karyotype. I believe that the choice should remain an individual option, based on receipt and consideration of accurate risk figures.

The question sometimes arises as to whether we should use amniocentesis to help us abort carriers for autosomal or X-linked genetic diseases. While I would support the right of any couple to have any pregnancy terminated, I think that the attempted eradication of asymptomatic carriers is based upon specious reasoning. For one thing, it is estimated that all of us carry three to five mutated, defective genes, capable of causing disease in a child conceived by a "wrong" mate. So, a strong effort to wipe out "bad genes" would lead to its being pretty lonely here on earth. Furthermore, genes known to cause disease in a double (homozygous) dose may actually be advantageous in the single (heterozygous) state. For example, a single dose of the recessive gene for sickle cell anemia confers upon its carrier a relative immunity to malaria. Moreover, as previously mentioned, despite the concern that present practice conditions may lead to an increase in the number of deleterious gene carriers in the population, it has been calculated that any such increase would be small, and most likely, of no practical significance. A request for abortion of a gene carrier is usually either based upon misinformation, or represents a conscious or a subconscious attempt to utilize a "socially acceptable reason" for termination of an unwanted pregnancy.

Sometimes, couples request amniocentesis but state that under no circumstance will they elect to abort the pregnancy. In this situation, most geneticists are reluctant to perform the procedure, since this would imply acceptance of the small but definite risk of amniocentesis without any hope of being able to use the information obtained to influence the management of the pregnancy. This objection is answered by the couples' assertions to the effect that prior knowledge of the presence of a birth defect would give them five months to adjust to the fact before the arrival of the baby. Considering both this attitude and the fact that most taps will provide reassuring information, it becomes difficult to deny amniocentesis to such patients, as long as they recognize and accept the intrinsic risks of the procedure.

Some patients request amniocentesis for sex determination, in order that they may abort a fetus of the undesired sex. Even though these requests are perfectly legal, very few geneticists are now honoring them. The reason usually given for refusal is that the laboratory facilities and manpower are not sufficient to handle the volume of work that would accrue if sex determination on request were to become routine, and that preference must be given to families seeking to avoid having children with genetic diseases. This is certainly true enough, but it would not be true to say that the attitude of the doctors is neutral regarding the idea. In fact, all geneticists with whom I've discussed the issue have expressed a negative opinion, ranging from mild distaste to frank revulsion. Eventually, however, I believe the point will become moot: it seems apparent that before laboratory space and technicians proliferate to the point where another excuse would have to be found, a reliable method of preconceptual sex determination will have been made available (see chapter 7). This will constitute an option far safer, and more palatable to both patients and doctors than postconceptual diagnosis and selective abortion.

Much has been made of the implications of the use of prenatal diagnosis in regard to the relationship of the individual to society. For example, it has been suggested that couples at risk for genetic disease in their offspring might be penalized by higher insurance rates if they will not have their pregnancies screened and selectively aborted. That may indeed come to pass, but what of it? Insurance companies routinely increase their rates to meet high risk situations. Persons with hypertension or diabetes pay higher life insurance premiums, but no one suggests that we should stop checking patients' blood pressures and blood sugars. To frame another analogy, it's quite lawful to build your house at a great distance from fire hydrants, but you should expect to pay for the privilege with increased fire insurance costs. And what of the couple who elect to use prenatal diagnosis? Since their children will be at lower risk for birth defects, they might qualify for reduced insurance premiums.

Ethicists have expressed the concern that the government eventually might make diagnostic amniocentesis mandatory, and then order abortion of all fetuses diagnosed as defective. I think this is unlikely. In our society, Jehovah's Witnesses are free to reject Rho-GAM, the medication that prevents Rh-incompatibility disease. Members of various cults may invoke their beliefs to avoid compulsory vaccinations. It seems much more likely that anti-abortionists will succeed in imposing their morality by fiat on pro-abortionists than the other way around. Furthermore, if our government does indeed reach the stage of being able to mandate abortion, then I suspect that will be one of our lesser problems.

Future Developments in Prenatal Diagnosis

Prenatal diagnosis of genetic disease has been in use for less than ten years. What future directions may the specialty take?

Several groups of investigators are now trying to devise improved methods of cell culture, with the aim of shortening the time needed to grow enough amniotic cells to perform chromosomal and biochemical analyses. In addition, micro-techniques for biochemical analyses are being developed so that fewer cells will be needed for assay. In these ways, it is hoped that the three- to five-week period of waiting for results may be shortened.

An exciting possibility is computer analysis of chromosomes. Current research indicates that a computer can be used to rapidly scan chromosomes on a slide, select suitable groups, and then perform accurate analysis. Since it takes a technician the better part of a day to prepare a report on one patient, and a computer can provide information on fifty or more patients per day, this means that the technology may soon be at hand to permit routine chromosomal screening of all pregnancies in women over thirty-five. In fact, since a 1970 survey of newborn infants revealed a chromosomal abnormality in 0.5 percent, screening might eventually be offered to all pregnant women. However, this latter possibility would also depend upon further information regarding the intrinsic risks of the tap and upon more definitive data evaluating whether some of the newly discovered chromosomal anomalies are associated with clinical abnormalities or whether they are harmless structural variants.

Hopefully, future developments will include methods for treating genetic diseases in the intrauterine patient, so that abortion will become unnecessary. Most likely, cures for chromosomal diseases will not be forthcoming within the foreseeable future, if ever. A tremendous quantity of genetic material is duplicated or lost when whole chromosomes or visible fragments are present in excess or are lacking. It would be a formidable task to alter the effects of multiple extra or deficient genes. Much more likely is the treatment of inborn errors of metabolism, diseases resulting from the dysfunction or nonfunction of a single gene pair. Here, the aim is to circumvent the affected metabolic pathway (as in the dietary therapy of galactosemia), to provide a necessary missing body chemical (as in congenital adrenal hyperplasia, an inborn error of metabolism where the adrenal glands are unable to form cortisone), or to administer the missing enzyme or even the missing gene itself (see previous chapter).

Within a few years, amniocentesis may be used to diagnose *in utero* infections that may lead to congenital anomalies. Organisms such as rubella (German measles) virus, cytomegalovirus, and the protozoan Toxoplasma all can produce physical defects and brain damage in the fetus. But even with the help of blood tests, it is frequently difficult to tell whether a maternal infection has crossed the placenta and reached the fetus. A group from Harvard Medical School has resolved such a dilemma for a patient by culturing rubella virus from her fetus' amniotic fluid. The aborted fetus was later shown to have tissue pathology characteristic of rubella infection. Further work along this line might lead to tests to identify virus or characteristic viral damage in the amniotic cells, thereby bringing the diagnosis one step closer to the patient.

A good deal of excitement has been generated over the experimental use of the fetoscope (also called amnioscope). This is a long, thin instrument used to visualize the fetus via a fiberoptic lens system. To this date, the smallest diameter scope designed has been two millimeters. In comparison, the diameter of the needle used to remove amniotic fluid is about one-half of one millimeter. Apparently, this is a crucial difference, because in experimental use of the fetoscope in women scheduled to undergo induced abortion, the death of the fetus has occurred in a large proportion of cases. Therefore, it is obvious that the technological problems are far from solved. At this time, an instrument small enough in diameter to compare with an amniocentesis needle would not permit anything near adequate visualization.

But this is not the only problem associated with fetoscopy. A couple of years ago, the procedure was being highly touted as a technique that might permit visual screening of all pregnancies for structural birth defects. Dr. Carlo Valenti of Brooklyn and Dr. J. B. Scrimgeour of Edinburgh enthusiastically reported having seen different parts of the fetal anatomy, and backed up their claims with many fine photographs. But recently, these and other fiberoptic pioneers, such as Dr. C. R. Wheeless of Johns Hopkins, have had to temper their enthusiasm. For one thing, there has been the problem of fetal demise after the procedure. But just as disheartening have been five areas of difficulty related to visualization. First, the narrow gauge of the instrument makes for a limited sighting angle. Second, the fetoscope must be very close to a fetal structure before it can be seen. Third, the rigidity of the scope and the limited intrauterine space make manipulations difficult. Fourth, leakage of only a few drops of blood into the amniotic fluid will render this liquid totally opaque. Fifth, unless the fetus can be made to rotate slowly around its long axis, the side turned away from the observer will remain as the dark side of the moon. That this is not a minor problem is illustrated by Dr. Scrimgeour's report of a case in which he had some difficulty in seeing all aspects of the fetus. No abnormality was evident, but on subsequent delivery of the fetus, a small spina bifida and hydrocephalus were found.

Even though it seems unlikely that fetoscopy will soon, if ever, attain wide use as a screening procedure, it could become of great value in some specific situations. If a scope of safe dimensions could be designed, one which would provide sufficient visualization to even dimly delineate the placenta and fetal extremities, the door would open to the prenatal diagnosis of a large number of new diseases. For example, a small sample of blood might be taken from a placental vessel. Since this would be fetal blood, many serious blood diseases could be diagnosed, the most prominent being sickle cell anemia. In fact, sickle hemoglobin already has been identified in as little as one-tenth of a milliliter of fetal blood obtained at abortion. Hemophilia could be diagnosed by assaying fetal blood for Factor VIII, the clotting factor missing in this disease. All that remains is to find a safe technique whereby such samples could be drawn. Furthermore, if a small muscle biopsy could be taken from a fetus at risk for muscular dystrophy, this other major X-linked disease might also become amenable to specific diagnosis. Muscle tissue from even young patients with muscular dystrophy shows characteristic changes that can be identified microscopically. Though not proven, it's likely that the same changes would appear in the muscles of an affected fetus.

At this time, the development of fetoscopy is being greatly hampered by recent congressional action which discourages the performance of research on live fetuses scheduled for abortion. These legislative bans are the result of the many-pronged attack of the anti-abortion movement, which aims to curtail all practices related to termination of pregnancy. Therefore, it may be many years before fetoscopy can attain practical value.

Probably the most promising development in prenatal diagnosis involves the use of ultrasound. As previously described, this technique currently is being used to localize the placenta and to help diagnose anencephaly. However, the work of Dr. Hugh Robinson of Glasgow makes it obvious that this is really just a beginning. Ordinary ultrasound prints pictures that show interfaces as white lines on a black background. But Dr. Robinson's experimental ultrasound machine operates in "grey scale," which means basically that the instrument is capable of recognizing and recording interfaces far subtler than hitherto possible. For example, conventional ultrasound machines will recognize and record the outlines of the skull bones and the midline structures of the brain. Refined grey scale techniques can reveal not only these, but also the scalp, the brain tissue itself, and the inner brain cavities which contain the cerebrospinal fluid. Dr. Robinson has observed the beating heart of a six-week-old fetus. He has "visualized" liver, kidneys, digestive tract, and most other internal organs.

So here we have a stupendous opportunity for the possible screening of all pregnancies for structural birth defects, utilizing a procedure that is painless, technically not difficult, and apparently safe. In addition, this technique would pick up the severe anomalies of internal organs that would be hidden from the most efficient fetoscope. If grey-scale ultrasound continues to fulfill its present degree of promise, and if further follow-up studies indicate the safety of the technique, in a few years we may have at our disposal a means for preventing birth defects in numbers that stagger the medical mind. We may well be on the threshold of a true breakthrough in this specialty.

Thus, prenatal diagnosis, the first significant means of therapeutically applying the information supplied by genetic counseling, appears to be the field that will for a long time continue to be at the forefront of applied genetic practice. Not until widespread genetic therapy becomes available is there likely to be a challenger to its preeminent position.

Part II

Reproductive Engineering

6

Artificial Insemination

Reproductive engineering is a field which includes any attempt to produce new individuals by means other than uncomplicated heterosexual intercourse. "Reproductive engineering" seems an appropriate term, since the use of novel techniques of reproduction involves making practical application of the knowledge of two pure sciences, genetics and reproductive biology.

In the next six chapters I will discuss reproductive engineering, proceeding from those concepts with the most nearly immediate application to the most futuristic. Artificial insemination, the subject of this chapter, is the only mode of reproductive engineering whose use is now possible at any level other than the experimental. In fact, artificial insemination has been with us for many years. During this time, it has provoked a splendid ongoing demonstration of how discussions of modifications in human reproductive practices tend to generate more heat than light. In this respect, artificial insemination has made predictable the present growing controversies over *in vitro* fertilization (chapter 8) and cloning (chapter 10).

The material in this chapter will be almost completely limited to the subject of human artificial insemination with sperm from a donor, or AID. Artificial insemination using the husband's semen (AIH) can also be performed, and may properly be considered a type of reproductive engineering. However, the technique utilized in AIH does not differ from that in AID, and there exists little controversy concerning the use of AIH. Therefore, it should suffice to say that AIH may be helpful when male or female anatomic abnormalities preclude natural insemination, or when a husband's ejaculates contain small numbers of sperm, so that fertilizability may be enhanced by pooling and concentrating several specimens.

History and Technology of AID in Humans

In the United States, donor insemination probably was first performed by Dr. Robert L. Dickinson, in 1890. By the 1930s its use had increased dramatically, largely due to the pioneering efforts of such physicians as Dr. Alan Guttmacher and Dr. Sophia Kleegman. There exist no accurate records of the use rate, but it has been estimated that between 5,000 and 10,000 AID babies are born in America each year.

The indications for the use of AID fall into two major categories. The more common indication for the procedure is relief of infertility in couples where the husband does not produce functional sperm. Here, the therapeutic desideratum does not primarily involve genetic considerations. However, the second indication for AID is to provide a reproductive alternative for couples in whom fertilization by the male consort is likely to produce a child with a serious genetic disease. For example, the husband might be Rh positive, and the wife Rh negative and sensitized; in this situation, AID could be performed with semen from an Rh negative donor. Alternatively, the husband might suffer from an autosomal dominant disease, carrying a 50 percent risk for the offspring; or both husband and wife might be heterozygous carriers for an autosomal recessive disorder which cannot be diagnosed prenatally. In either of these two situations, AID would greatly reduce the risk of genetic disease in the child.

The AID procedure should begin with a lengthy interview session (or sessions) in which the technique is thoroughly explained to the couple, and at which time the counselor assures himself that both spouses truly desire to utilize the procedure. He should be certain that there is no ambivalence, and that neither partner has been coerced. Then, it should be ascertained that the woman is indeed fertile: she should ovulate properly, and there must be no mechanical obstruction to the union of sperm and egg, such as blockage of the fallopian tubes. It is terribly embarrassing for a doctor to discover that he has been inseminating a woman who is incapable of being fertilized.

Next, the physician chooses a donor, being careful to maintain complete anonymity between donor and recipient couple. Donor semen is obtained by masturbation into a dry, clean receptacle. Since the initial portion of an ejaculate usually contains the highest number of fertile sperm, some inseminators obtain a fractional sample, and use only the first part. Others, however, inseminate with the entire sample. Although most writers in the field stress that they attempt to provide a donor free of "genetic taint," we can infer from the previous chapters how difficult a goal this is, even assuming good mental and physical health, and a spotless family history. In addition, most experts insist upon compatibility between blood types of donor male and recipient female; many also try to match the physical appearance of the donor and the husband. Most donors are paid between $25 and $50 for their services. Because of convenience, a large proportion of donors are selected from the ranks of medical students, interns, and resident physicians.

In recent years, some artificial inseminators have been utilizing previously frozen, stored semen. This aspect of AID will be discussed later in the chapter.

The actual insemination procedure is simple. One of three techniques is used (Fig. 6.1). A small volume of semen may be introduced into the cervix (intravaginal extension of the uterus); or, the donor ejaculate may be placed at the top of the vagina, so that it can bathe the cervix, as would occur after natural insemination; or, a plastic cap may be applied over the cervix, the semen introduced inside the cap, and the cap left over the cervix for twenty-four hours. Semen usually is not introduced within the uterine cavity, because some of the normal vaginal bacteria may be carried along, producing infection; furthermore, seminal fluid may irritate the uterus, causing it to contract painfully.

The particular technique used in artificial insemination does not seem to significantly affect the success rate. When fresh semen is used, approximately 35 percent of women conceive during the first cycle, and 75 percent by the fourth attempt. When frozen semen is used, an eventual 50 percent rate of success is usual.

Insemination customarily is performed around the presumed time of ovulation in the recipient. The time of ovulation may be estimated by any of several biological parameters, the most frequently used of which is a minor but constant elevation in body temperature. Most doctors perform between one and four inseminations near the time of ovulation. Sometimes the semen of the husband is mixed with that of the donor, or the couple is asked to have intercourse after AID has been performed. These maneuvers are undertaken both to make the husband feel less left out of things, and to introduce the theoretical possibility that the husband just might become the biological father. Dr. Guttmacher, however, did not approve of this practice, referring to it as an immature form of subterfuge.

Three of Dr. Guttmacher's general principles of practice for AID are worthy of quotation and comment. First, he emphasized most strongly that AID must never be urged upon a reluctant or ambivalent couple, that this would inevitably lead to emotional and psychological difficulties. Second, he did not have his patients sign the lengthy consent forms which are a part of standard practice; he felt that this forced the couple to remember what should be forgotten. In his opinion, avoidance of written consent helped the husbands to better

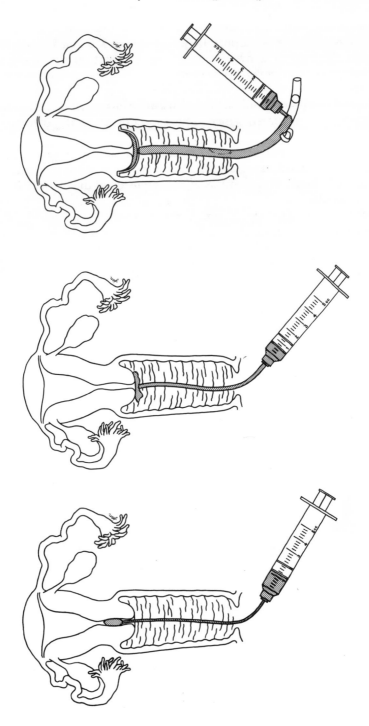

Fig. 6.1:   The three general techniques for artificial insemination in humans.

accept the children as their own, right from the time of birth. Third, he assigned paternity to the husbands on the birth certificates. The questionable legality of this practice will be discussed shortly. To avoid problems, some inseminators recommend that the physician who performs AID should not be the one who delivers the baby, and that the delivering obstetrician should be kept in ignorance of the AID. Dr. Guttmacher wrote that his only concern was to do the best thing for the patient and that the "legal lie" didn't bother him. (Those readers inclined to take exception to Dr. Guttmacher's choice of active rather than passive dissimulation are referred to "On the Decay of the Art of Lying," a pertinent essay by Mark Twain.)[1]

Problems Associated with AID

Moral and Ethical Problems

The use of AID could not be justified if children conceived by this means were at excess risk of suffering birth defects. Because of the secrecy that surrounds the procedure, such statistics are not easy to come by, but the few studies done to date seem to indicate that no problem exists in this area.

On the emotional level, AID has never met with overwhelming acceptance. In reference to the frequent strong opposition to the procedure, Dr. Sophia Kleegman wrote:

> In any emotionally charged area, a change in mores must be preceded by a change in practice. All through the ages the sequence has been the same: Rigid negation; negation with diminishing horror; gradual curiosity, study, and evaluation; slow but steady acceptance.[2]

Most of American society now seems to be at Kleegman's second or third level. However, several religions hold that AID is invariably immoral and indefensible: these faiths include Orthodox Judaism, Lutheranism, and Roman Catholicism. Beginning with Leo in 1897, Popes have repeatedly spoken against artificial insemination. Some of the Catholic opposition is based on the premises that AID is an "unnatural act," and tends to "lower man to the level of animals." Catholics also object to AID because the procedure involves both masturbation and (in their opinion) adultery. The adultery issue is interesting: although most persons and lexicographers hold that the essence of adultery is the sexual act, conservative clergymen and even some courts of law have held that AID is indeed adulterous. That we frequently are not totally rational beings is borne out by the statement of many men to the effect that they do not consider AID adultery, but would nonetheless look with repugnance on the fertilization of their wives by the sperm of other men. (In fairness, the great majority of these men also say that they would not wish their sperm to be used to fertilize other men's wives.) Many women, too, wish to maintain bilaterally exclusive reproductive privileges with their husbands.

Excellent presentations of religious attitudes toward AID can be found in the articles by Guttmacher and Weinstock (see references).

The secrecy involved in AID is a major issue; it cannot help but reinforce the thought that something not quite proper is being indulged in. Nothing done behind a curtain of deception and denial can ever be looked on as other than improper, and so, I don't believe that AID can hope to achieve respectability until a number of courageous patients and doctors lead the way by going through the procedure in an open fashion. Formerly, the common belief was that adopted children should not be told they were adopted; now, these children are dealt with in a more honest manner. As the natural parents of adopted children are kept anonymous, so the identity of sperm donors can still be maintained in confidentiality.

Questions of ethics and morality have been raised with regard to semen donors. For one thing, the fact that these individuals are paid may lead to problems analogous to those that have been encountered with paid blood donors: to be certain that they will not be rejected, prospective semen donors might falsify their medical or genetic histories. Furthermore, some psychiatrists have wondered in print about which sort of individual it is who obtains gratification from semen donation. Is he an unselfish, loving person, happy in the knowledge that he is helping an infertile couple to have a child? Or is he an egomaniacal narcissist? Is it commendable or even desirable that a man should father a child, and never wonder about his offspring?

Most writers in the field state that AID should be offered only to married couples, specifically declaring that single women should never be inseminated. This point deserves discussion. Today, single persons are adopting babies, and seem to be doing a good job of raising them. One might well ask why a single woman should be categorically ineligible for AID: this procedure would enable her to bear and raise her own child, without having to become sexually involved with a man in a degrading stud relationship. In fact, AID for single women might well offer a convenient mechanism by which artificial insemination could be brought out into the open.

The concern has been expressed that increased use of AID might lead to a major and serious increase in unwitting incestuous marriages. As long as one donor is not used to tremendous excess, this possibility seems remote. Certainly, adoption offers the same potentiality, but no one seems to lose any sleep over this. Moreover, many geneticists believe that the rate of fraudulent paternity in the offspring of married couples is as high as 5 percent. A blood-type survey in a small village in southeastern England revealed that fully 30 percent of the children could not have been fathered by their presumptive fathers. In the face of figures such as these, AID would seem to pose no great problem.

Psychological Problems

This aspect of AID also suffers from severe insufficiency of hard data. Most of the available information is anecdotal. By and large, proponents of AID paint pictures of happy, grateful parents and rosy-cheeked children, while opponents describe dreadful maladjustments in everyone concerned.

Drs. Guttmacher and Kleegman, and S. J. Behrman have cited divorce rates in their patients as being about twenty times lower than the rate in the general population. In addition, they have offered the opinion that their patients have suffered no increase in neurotic or psychotic mental diseases. Their findings suggest that careful screening of prospective couples by gynecologist-inseminators will result in limitation of the procedure to well-adjusted persons.

This approach seems to be indirectly supported by the psychiatrist Dr. Herbert Peyser. He reported two detailed cases where AID had been urged by doctors upon infertile couples. In one case, the husband suffered serious mental decompensation; in the second case, the wife did. It seems clear that no couple or person should ever be pressured to undergo AID; in addition, it may well be that AID should not even be offered by the doctor, but only assented to after patient request, and subsequent thorough discussion.

On the other hand, it cannot be assumed that careful preinsemination screening will certainly protect against undue psychological complications. Peyser mentioned a British physician who gave up the practice of AID because of what he felt to be obvious feelings of inferiority on the part of the husbands and undesirable mental attitudes of the wives toward the anonymous donors.

In addition, the psychoanalyst Dr. Gerda Gerstel has reached strongly negative conclusions about AID. Apparently utilizing a strict Freudian approach, she reported five disastrous AID cases, all of which had been approved by "reliable and ethical gynecologists." Each of the husbands suffered severe depression, stemming originally from the diagnosis of azoospermia, and worsened by the decision to participate in AID. Dr. Gerstel believed that the husbands perceived the donors as men with bigger penises than their own: in other words, their fathers. This, then, led to self-directed hostility, and eventual depression. The wives suffered disturbing sexual fantasies about the donors, and sometimes about the doctors. They rejected the babies prior to birth, postponing the wearing of maternity clothing as long as possible. The maternal rejection continued after the babies were born. As might be expected, all the children suffered severe emotional damage, featuring separation anxiety, and faulty identification and maturation. On interview, each child was found to be aware of a "dark family secret" involving the father. Dr. Gerstel concluded that upon learning of his infertility, a man will inevitably suffer some degree of neurotic reaction, and that AID will almost certainly worsen his symptoms. Her final statement was that in her opinion, the decision to utilize AID might be considered to indicate the presence of an emotional disturbance.

At this time, Freudian analysis is largely out of fashion, and there may be a tendency to pooh-pooh Dr. Gerstel's interpretations and conclusions. However, the fact remains that she did present several cases where AID, by whatever mechanism, produced a terrible outcome for all three people involved: husband, wife, and child. Contrariwise, the aforementioned proponents of AID can offer reassuring data. It will take a good deal of careful investigation by unbiased observers (if there can be such a thing) to reveal the true situation with regard to the psychological complications of AID, and how to avoid them.

Legal Problems

As a basically conservative institution, our legal and judicial system has encountered considerable difficulty in trying to come to grips with AID. The problems posed by different aspects of the procedure are entirely novel, and so, without precedent. Therefore, the current status of AID in the eyes of the law is characterized by much confusion and disagreement. Although the *Yale Law Journal* has referred to AID as "a parvenu in the field of law," the fault cannot be shifted to AID itself, even by such adroit verbal gymnastics. The true problem lies in the difficulty of making our legal machinery more responsive to rapid change.

The English are no better off than we are. Under British law, children resulting from the use of AID are not recognized as "lawful," or legitimate. This decision is primarily based on the adultery analogy.

The first North American legal case involving AID occurred in 1921, when an Ontario court formulated the opinion that the use of AID constituted adultery. Part of this decision included the comment that Moses would have looked with repugnance upon the procedure.

The first case in the United States was a divorce proceeding in Illinois in 1948. Although the judge held that AID could not be considered adultery, he still granted the divorce to the husband, since the wife had also engaged in conventional adulterous undertakings. Another 1948 ruling, this one in a New York court, also held that AID was not a form of adultery, and further specified that the child was to be considered legitimate, as long as the husband had given his consent to the procedure.

Reason received a setback, however, in the 1954 *Doornbos* vs. *Doornbos* decision in Illinois. Here, it was decided that AID was contrary to public policy and good morals, and did indeed constitute adultery. Whether or not the husband gave consent was considered immaterial. Thus, the child was labeled illegitimate, and the husband was said to have no legal right or interest in the child. On the other hand, in 1968, a court in California ruled an AID child legitimate, and declared that not only did the husband have legal rights and interests, but legal obligations, as well. He was ordered to pay child support. A similar attitude was taken by a New York judge in 1973, who refused to allow a woman's second husband to adopt her AID child without the consent of the first husband; the first husband was held to have all the rights of a "natural father" with respect to the child.

A Uniform Parentage Act, drafted in 1973, included a section on AID, in which attempts were made to resolve some of these inconsistencies. Basically it was proposed that the consenting husband automatically be regarded as the legal father of the child, and that the semen donor have no legal relationship whatever to his biological offspring. Although some legal experts have expressed concern over semantic imprecisions in the draft, they still hope that the act eventually may bring some order out of the legal chaos. In the meanwhile, California, Georgia, Oklahoma, and Kansas have passed laws officially legalizing AID and legitimizing the offspring, provided the husband had consented to the procedure.

It has been hypothesized that the doctor who performs AID may be at legal risk in several respects. He might be sued if a defective child is born. He could be indicted for perjury if he knowingly attributed paternity to the husband on the child's birth certificate. He might be cited as a co-conspirator in adultery, or, if the wife is under legal age, he might even be charged with rape. Finally, a wife might decide to seek child support from him. Some of these situations do sound farfetched, but a judge or a jury with strong negative feelings toward AID might well decide against the physician. This would be especially likely in a case involving the birth of a defective child. Although no legal action for AID-related damages has yet been taken against a doctor, virtually all professional artificial inseminators strongly advise that a full consent form be signed by both members of the recipient couple, as well as by the donor and his wife. A doctor still could be damned even if he did utilize such a form, but he'd certainly be doubly damned if he didn't.

Although, to date, no donors have been involved in legal maneuvers either, there is concern for their situation as well. Could anonymity be breached, either surreptitiously or by court order, so that the wife could sue the donor for child support? Could the child claim a portion of the inheritance from the estate of its donor-father? Is it possible that an occasional donor, unable to psychologically relinquish the child, might attempt to gain legal custody? Again, none of these situations is terribly likely to happen, but they do not appear to be beyond the realm of possibility.

For more than fifty years, our courts have demonstrated uncertainty in dealing with many aspects of AID. Fortunately, a number of lawyers and law professors are now taking an active interest in coping with problems related to reproductive engineering; therefore, it is to be hoped that the legal future of this field will be more reasonable, consistent, and humane than has been its past.

AID in the Future

Frozen Semen Banking; Frozen Egg Storage
(Cryobanking of Sperm and Eggs)

For many years, frozen semen has been used routinely for artificial insemination in cattle. Semen can be obtained from bulls and kept in deep freeze for ten years or longer without significant loss in fertilizing potential. Experiments have demonstrated that there exist marked species differences with regard to sperm survival after freezing and thawing. Bull sperm maintain high fertilizing capacity, while ram sperm lose most of their ability to fertilize eggs. Frozen boar sperm lose all potential for fertilization. In contrast, the ability of human sperm to withstand freezing appears to be high. This basic fact underlies the current interest in banking human semen for research and commercial uses.

The practical use of frozen sperm began in 1949, when Dr. C. Polge of England made the discovery that a strong cryoprotective effect could be obtained by adding a small volume of glycerol to the semen before freezing. All previous freezing techniques produced unacceptable damage to the sperm. It is considered that glycerol may produce its beneficial actions during the freezing process through protection of the proteins and inorganic ions in the sperm. Some workers now use a cryoprotective mixture of glycerol and egg yolk; the latter substance also seems to protect cellular proteins.

The general procedure for freezing sperm is as follows: a small quantity of an ejaculate is initially checked for total number of sperm present, percent of normally motile cells, and percent of cells of normal morphology. Then, the remainder of the semen is mixed in a 12:1 ratio with the cryoprotective agent, and the fluid is divided into 1 ml.

portions. Each portion is placed in a glass or plastic vial. Next, the containers are cooled at a slow rate (one degree per minute) from room temperature to the freezing level of 0°C. More rapid cooling above the freezing level causes large numbers of the sperm to lose their functional capacity, a phenomenon known as cold shock or temperature shock. Between 0°C and -30°C, the semen is cooled at the more rapid rate of five to seven degrees per minute. This is the temperature range at which crystals form, and prolonged exposure to the consequent heat of crystallization would also injure the sperm. Below -30°C, the semen may be cooled rapidly, and when the temperature reaches -80°C, the vials are placed in a liquid-nitrogen freezer maintained at -196°C. Thawing is less complicated; it is usually done by placing the sample in air at room temperature, or in an alcohol bath at 5°C.

Dr. J. K. Sherman has proposed several possible applications for semen banking. These include (1) collection and pooling of several ejaculates from a husband who produces subfertile numbers of sperm, (2) the possibility of multiple inseminations per cycle with donor semen, thereby reducing the number of cycles necessary to achieve fertilization, (3) providing a more extensive choice of donors, in order to facilitate matches in blood groups and other desired traits, (4) "reproductive insurance," for men about to undergo vasectomy, and (5) eutelegenesis (see below). Dr. Sherman believes that the present state of our knowledge justifies the use of frozen human semen "in suitable clinical situations." However, not everyone would agree with him.

To begin with, it would be desirable to have more knowledge regarding the effects produced on sperm by storage at low temperatures. Which cellular metabolic functions are altered by low temperatures, and in what fashion? Which effects are permanent, and which reversible? A variable proportion of sperm in different samples are killed by freezing: what are the factors that favor survival? In truth, our knowledge of sperm cryobiology is extremely limited.

It is assumed that the only part of the spermatozoan which survives after fertilization is the hereditary material, the chromatin. All the necessary embryonic cytoplasmic structures are thought to originate from the egg. Therefore, it would be most important to determine whether freezing sperm induces gene mutations. A tenfold increase in the spontaneous mutation rate would certainly be significant, but it would take millions of closely followed inseminations to uncover such an occurrence. This is so because the spontaneous mutation rate is of a low order of magnitude, and because only dominant (not recessive) mutations would be apparent in the original mutants. This kind of information might be gathered more expeditiously by the use of frozen-semen insemination in highly inbred strains of mice, which are genetically homogeneous. In such an animal population, a large proportion of any mutations that were to occur would be apparent in the first and second generations after insemination. However, no such research has been tried.

To date, the world literature contains reports of fewer than 1,000 conceptions after insemination with frozen semen. It should be emphasized that in the overwhelming majority of these cases, the semen had been stored for under three years, and most often, for less than six months. Only a few successful pregnancies have resulted after the use of semen stored ten years or longer. The collected data indicate that the use of frozen semen is not associated with increased rates of miscarriage or birth defects. However, a disturbingly large number of the reports are more than a little vague with regard to the actual details of the follow-up procedures.

Probably the best reports are those of Drs. Keith Smith and Emil Steinberger of Houston, who uncovered a good deal of interesting data. The conception rate in their patients was 61 percent with frozen semen, in comparison to 73 percent when fresh semen was used. Furthermore, after insemination with fresh semen, most conceptions occurred in the first or second cycles, but it took an average of three to four cycles to achieve success with frozen semen. And when conception did occur with frozen semen, it was discovered that the successful sperm samples had had superior prefreezing numerical counts and motility assays. This fact led the authors to suggest that perhaps only semen of high quality should be considered suitable for long-term freezing.

Smith and Steinberger also reported that immediate freezing and thawing would invariably produce a significant loss in sperm motility. Cryostorage for up to thirty-six months resulted in no greater impairment, but from that point on, motility progressively worsened, until at five years, only about 20 percent of the original motility remained. This fact contrasted with Dr. Sherman's data which showed no significant drop in motility after even ten years of storage. Obviously, questions related to individual variations in freezing and storage techniques remain to be answered.

Despite the paucity of experimental and clinical data concerning the freezing and storage of human semen, the past few years have seen a proliferation of cryobanking facilities. By late 1973, there were sixteen semen banks in the United States, including three commercial organizations with seven branch offices. It appears that two basic considerations underlie the establishment of these banks: an increased demand for AID by infertile couples, and the desire for "reproductive insurance" by large numbers of men, most of whom wish to use vasectomy as a means of fertility control.

Even so strong an advocate of frozen semen as Dr. Sherman is made uneasy by the potential for harm that seems inherent in commercial semen banks, as they are now being run.

Semen banks are under no form of control or regulation. Apparently, no consistent mechanism exists to prevent the unauthorized use of a depositor's banked sperm for AID situations, or in research procedures such as *in vitro* fertilization experiments. There exist no standard forms or procedures for providing informed consent. But, highly informed consent would seem most desirable, both for the donor and for the recipient of frozen semen.

For example, consider the man about to undergo vasectomy. Publicity articles and advertisements may well lead him to believe that frozen semen can provide reliable "reproductive insurance." But no one really knows how long sperm can be safely stored in the frozen state. Moreover, as we have seen, there may be an inevitable decline in the fertilization capacity once sperm have been frozen; with poorer samples, this impairment appears to be considerable. Furthermore, nothing is known about the genetic changes that freezing temperatures may induce in sperm. All things taken together, I think Lloyd's would put a rather high premium on this kind of insurance.

The recipient of frozen semen should have some concerns as well. Aside from the questions about the genetic safety of frozen sperm, there may be specific donor-related problems. We know that many persons have contracted diseases after being transfused with blood purchased by blood banks from derelicts eager to give up a pint of blood for the price of a few bottles of wine or a supply of heroin. Can we be certain that commercial semen banks will be more judicious in their choice of donors? For example, gonorrhea has already been reported to have been transmitted by AID. This issue brings up the whole question of whether semen donors should ever be paid. Again, if we can analogize from the blood situation, it may well be that the answer is no — especially if we stop to realize that banked semen is much less likely than blood to be needed for relief of a medical emergency.

With fewer than 1,000 AID babies having been born after the use of frozen semen, it's difficult to accept the claim that this should be considered a standard procedure. Women scheduled for insemination with frozen semen should be told that the risks are far from fully elucidated. This, in turn, brings up another point: in all existing reports of the clinical use of frozen semen, no statement has ever been made that the recipients were told they were engaged in an experimental undertaking, or that they had furnished informed consent.

Semen cryobanking is showing every sign of becoming a popular fad; it may well persist until a number of unfortunate results force it out of fashion.

There has been some discussion of a companion process to semen cryobanking: the frozen storage of human eggs, for use in artificial inovulation (analogous to artificial insemination). Such a concept would find application in women who are infertile either because their ovaries contain no eggs (as in XO Turner's syndrome or premature menopause) or because their ovaries have been removed. Artificial inovulation might also be applied for eutelegenetic purposes (see below). Despite the fact that the technical problems associated with the freezing of human eggs should not present insuperable obstacles, it isn't likely that artificial inovulation will ever achieve wide use. For one thing, although a man can produce limitless numbers of sperm, only a small and finite number of eggs can ever be extracted from an ovary. Furthermore, eggs cannot be obtained without surgically entering the female abdomen; similarly, an ovum recipient would have to undergo a surgical procedure. Therefore, *in vitro* fertilization of either a fresh or a frozen egg (chapter 8) with subsequent implantation of the developing embryo, would seem to be a more logical procedure than artificial inovulation, where, after all the work, it would not even be certain that the egg would be fertilized.

Eutelegenesis

Writing in the *Eugenics Review* in 1935, Herbert Brewer proposed that telegenesis be defined as "the process of reproduction from the germ cells of individuals between whom is no bodily contact. The possible application of this process to the eugenic breeding of man may be termed *eutelegenesis*."[3]

By the application of eutelegenetic principles and techniques, Brewer hoped to further the creation of excellence which he considered to be the "mission" of eugenics. Since it was known even then that low temperatures lengthened the fertilizing life of animal sperm, Brewer suggested that this phenomenon be combined with dilution of ejaculates to make possible the fertilization of many women with the sperm of a "few superlative individuals," who would be chosen by the usual group of eugenical experts. Brewer felt that most men would not object to this concept, since "there is no dishonor or humiliation in accepting, in common with the multitude of men, the position of second best to the ideal."[4] Brewer's suggestions, however, never were placed into practice, because within a few years, the Nazi atrocities had put all of American eugenics into cold storage.

In the late 1940s, the concept of eutelegenesis was revived, largely due to the efforts of Dr. H. J. Muller. In his desire to avoid a "genetic twilight," he thought that eutelegenesis utilizing frozen semen could function as an "entering wedge" for the eventual general acceptance of his eugenic recommendations. Presuming that couples using AID for infertility would want their offspring to have the best possible genetic endowment, Muller suggested the routine use of frozen semen in this circumstance. He felt that when it became apparent how superior these children would be, no reasonable person would want to reproduce by any other means. Muller, too, believed that the average man would not object to being reproductively preempted; in fact, he urged that the identity of the donors not be kept secret. Great scientist that he was, Muller's enthusiasm sometimes outweighed his objectivity, for example, when he said, "deep-frozen sperm . . . can be kept in perfect condition for an indefinite length of time,"[5] a comment which no cryobiologist would dream of making, even today.

Many objections can be raised to eutelegenesis, aside from those simply related to the use of frozen semen. Mostly, the objections have to do with potential emotional and psychological difficulties. For one thing, it's highly unlikely that many husbands would cheerfully step aside, adjudging their own traits less desirable of perpetuation than those of a "superior" donor. The severe types of psychological reactions that might be caused by eutelegenesis are illustrated by the previously cited cases of Peyser and Gerstel. Furthermore, if most men did not especially care to father their own children, there hardly would be such a call as there is now for prevasectomy "reproductive insurance."

Then, there are questions concerning the children. How would they interact with their "inferior" legal fathers? Would relative "superiority" of one's biological father become a new status symbol? What of the child whose donor-father fell from favor? Alas for Young Stalin in Russia and Young Nixon in America! What would happen to children who didn't measure up to the greatness of their biological fathers? After all, genius is a fragile commodity, and the children would share only half the genes of the donor, and none of his environment. In fact, if the donor were still alive, might he be sufficiently egomaniacal to resent the "underachievement" of one of his offspring, and try to gain custody of the child, claiming that the legal parents were incapable of properly nurturing a young genius?

With regard to donor selection, it would be more than a little difficult to find men who have only good traits to pass on: the personal and even professional lives of many geniuses have been unalloyed disasters. Furthermore, since it is usually young men who become natural fathers, how can anyone predict with any degree of accuracy which young men should have their own children and which should not? And since it is easier to evaluate the achievements of old rather than young men, what is to be done about the generally lowered fertilizing capacity of sperm and the increased likelihood of genetic mutations in oldsters who have been selected as donors?

By making possible large numbers of offspring from one donor, frozen-semen eutelegenesis would diminish our genetic diversity, thereby depriving us of one of our potentially most useful hedges against the uncertainties of the future.

Finally, the calculations of John Maynard Smith indicated that, in any case, eutelegenesis is unlikely to accomplish much in the way of desired results. Even if as many as one percent of all women were to have a child by eutelegenesis, the average IQ of the general population would rise only 0.04 points per generation. And even if a small, isolated group of people were to rely on eutelegenesis for reproduction, the average IQ of that population would rise only about fifteen points in 100 years. This is no more than could be obtained by beneficial environmental manipulations, and, as Smith said, "seems hardly sufficient to justify the establishment of a new religion."[6]

Conclusions

Artificial insemination with donor semen is the only technique of reproductive engineering now in general clinical use; however, it should not be thought of as a thoroughly established procedure. Although the technique is straightforward, many questions remain to be answered regarding the emotional and psychological aspects of AID, as concern husband, wife, child, and donor.

I can't feel any enthusiasm for the eugenic application of artificial insemination, but the use of AID seems reasonable at this time for properly screened and thoroughly informed individual couples who wish to employ it to alleviate male infertility or to avoid the birth of a child with a genetic disease.

I believe that the use of frozen semen should be considered experimental, and that semen cryobanks should be maintained only as research ventures. Before we have to worry about coming to grips with the problems of commercialization, we ought to answer some of the basic cryobiological questions. It must be kept in mind that there exist no compelling reasons to use frozen semen in a routine clinical fashion; convenience for the inseminator seems to be the outstanding advantage. Certainly, offering prevasectomy "reproductive insurance" is a dubious proposition. The World Planned Parenthood Association (hardly a reactionary organization) has taken a strong stance against this practice.

As each new concept of reproductive engineering becomes a technical possibility, the story of AID may be repeated. Many persons will scream for bans and plead for suppression, in the name of all that is right and proper. On the other hand, some laymen will clamor for immediate access to the fruits of scientific research, and a few doctors will be tempted to accede, hoping to scoop headlines for themselves. Perhaps, though, we will learn enough from history to avoid continually having to repeat it. It would be better if the march of genetic progress does not degenerate into a disorderly stampede.

7

Sex Determination

Since ancient times, prospective parents have wished for the ability to predetermine the sex of their children. Most attempts toward this end have been pure superstitions, always useless and often funny. For example, it has been said that conception of male fetuses bears an association with a north wind, or with a north-south orientation of the parents during mating. In addition, a full moon, the presence of an ax in the marital bed, or the wearing of boots or a hat during intercourse have all been touted as methods which predispose to the birth of a boy. Nor are such practices obsolete. I have taken care of patients who had been careful to have intercourse at the precise hour that their consulting astrologers had told them their stars would favor conception of a child of the desired sex.

Many superstitions concerning sex selection have been based upon right-left considerations. Usually, males have been associated with the right side, females with the left. For example, it has been suggested that a boy will be conceived if the woman lies on her right side after intercourse; or, a girl will result if the man hangs his trousers on the left bedpost. It even has been believed that eggs ovulated from the right ovary produce boys, while girls originate from left-sided ovulations.

A reliable method of sex determination could be of considerable importance. Some sociologists and demographers believe that much of our current population problem is related to dissatisfaction over family makeup, and that the ability to predetermine the sex of our offspring would result in generally smaller families. We might also cite the case of the former Queen of Iran, who was dethroned for presenting the Shah with an unending succession of girls.

In divorcing his wife, however, the Iranian Shah demonstrated more than callousness; he also showed the world his ignorance of basic biology. As mentioned in the Introduction, fetal sex is determined at the time of conception, when the twenty-three egg chromosomes recombine with the twenty-three sperm chromosomes. Normal eggs have twenty-two autosomes plus an X sex chromosome, while sperm have *either* an X *or* a Y to go with their twenty-two autosomes. Hence, it is the father who determines the sex of the child: an X-bearing sperm combines with the egg's X to give rise to a girl, while a Y-bearing sperm produces a boy, with an XY sex chromosome constitution. In sex determination, the "active factor" is the Y, since this chromosome appears to direct the sexually bipotential early embryo to differentiate along male anatomic and functional lines. In the absence of a Y chromosome, female differentiation occurs. Usually, this is in the presence of two X chromosomes, and, indeed, two normal X's appear to be a prerequisite to the development of a normally functioning pair of ovaries. However, patients with forty-five chromosomes including only one X (Turner's syndrome) will develop as anatomic females. So will the rare 46, XY individuals whose Y's are functionally defective. Although infertile due to ovarian nondevelopment, these persons have normal breasts, vaginas, and uteri, and, aside from infertility, function well in a female psychosexual role.

The Shah of Iran might argue that it is at least theoretically possible that the vaginal environment in some women is somehow inimical to the survival of one type of sperm or the other. However, at this time, no evidence exists to support such a concept, and human reproductive biologists universally hold that sex determination is a paternal function.

Sex ratio may be defined as 100 times the ratio of males to females in a given population. Human sex ratios vary markedly depending on the age of the population under consideration. Basically, at every stage of existence, male mortality exceeds that of the female. The true sex ratio at conception is not known, but more male than female fetuses are conceived. Estimates as high as 160/100 have been made, but these probably are excessively high, since they were based not on karyotyping, but on examination of the external genitals of early abortion specimens. At that stage of development, it is easy to mistake the female fetus for the male. Throughout pregnancy, an excessive loss of male fetuses by miscarriage results in a sex ratio at birth of 105 males to 100 females. The heightened mortality in males continues, and, by old age, the sex ratio becomes 70/100 or less.

Most people probably would be surprised to hear that there now exists a technique which would allow a couple to avoid having a child of the undesired sex. This procedure was discussed in chapter 5. It would involve the performance of amniocentesis at fifteen to sixteen weeks of pregnancy, with karyotyping of the fetal cells. Then, if the fetus were of the "wrong sex," abortion could be carried out.

This arrangement is both highly reliable and perfectly legal. In fact, it is the standard procedure for a couple at risk for an X-linked disease, such as hemophilia or muscular dystrophy. However, there are several reasons why it is not likely to achieve any degree of widespread acceptance in instances of simple sex preference. For one thing, facilities and personnel for the culture and karyotyping of amniotic-fluid cells are in short supply. Therefore, obstetricians and geneticists generally are reluctant to commit their capabilities to the attempted resolution of a problem which is by any standard considerably less compelling than the prenatal diagnosis and prevention of genetic diseases. Moreover, such an approach commits the couple to more than four months of pregnancy, with all the attendant emotional investment.

Abortion after sixteen weeks' gestation is not a trivial affair: it is performed by injection into the uterus of either a hormone (prostaglandin) solution or a solution of concentrated salt. If these injections fail to cause the uterus to empty, then a major surgical operation comparable to cesarean section must be performed. Furthermore, when used for sex determination, amniocentesis is a "half-way" procedure. It will not guarantee the birth of a child of the desired sex, but only the avoidance of the unwanted sex. It would be entirely possible for a women to have three, four, or even more abortions before coming up with a lucky throw of the genetic dice.

During the past few years, there has been some experimental interest in trying to analyze cells obtained from cervical mucus during the first three months of pregnancy. Such cells could originate from both the mother's uterus and the child's amniotic membrane. When stained by the quinacrine-mustard fluorescent technique, cells from a male fetus would demonstrate a Y-body (male sex chromatin). Total absence of Y-body positive cells in a sample would suggest the presence of a female fetus. Some researchers have claimed to be able to accurately predict fetal sex by this method, but others consider the technique to be unreliable. If, in time, fluorochrome staining of cells in cervical mucus does prove useful, it would permit fetal sex determination early enough in pregnancy to permit abortion by simple curettage. However, the objection would still hold that the technique would not permit a couple to be certain that they will, in fact, ever be able to have a child of the desired sex.

Therefore, it would appear that *postconceptual* methods of sex determination will always be poor expedients. Far more reasonable would be *preconceptual* techniques to permit not only avoidance of children of the "undesired" sex, but also highly reliable means to certify the conception of children of the "desired" sex. The remainder of the chapter will be concerned with preconceptual sex determination.

Potential mechanisms for preconceptual determination of a child's sex can be grouped into two general categories. First, there are the attempts to separate the husband's ejaculated sperm into X-bearing and Y-bearing groups, which would then permit sex choice by artificial insemination, using only the desired type of sperm. The second general category of maneuvers for preconceptual sex determination are more empirical, depending upon modifications of natural insemination in such a manner as to favor the survival or demise of one or the other type of sperm.

Separation of X-bearing and Y-bearing Sperm

Between the turn of the century and 1960, numerous attempts were made by scientists throughout the world to separate sperm by utilizing various physical and chemical differentiation techniques. Electrophoresis (like that used to separate mutant hemoglobins) was employed, in the belief that Y-bearing sperm would migrate toward the positive end of an electrical field, and X-bearing sperm toward the negative end. Unfortunately, experience did not bear out this idea. Similarly, early enthusiasm for the use of alkaline solutions to favor the survival of Y-bearing sperm faded under careful scrutiny. Then the realization that the X chromosome is much larger than the Y led to the demonstration that the set of twenty-three "female" chromosomes is about 4 percent heavier than the "male" array of twenty-two autosomes plus Y. Consequently, attempts were made to separate X-bearing from Y-bearing sperm by differential centrifugation and sedimentation procedures. However, these techniques never were terribly effective, and, in addition, high-speed centrifugation forces were often found to damage the sperm. In 1972, Dr. A. M. Roberts of London, in a highly technical paper, explained that there exist considerations inherent in the gravitational techniques which override the small weight differences of the sperm. Roberts offered the opinion that these attempts at separation would never be useful.

In 1960, the field of sperm separation was the scene of a lively little skirmish. The action began when Dr. Landrum Shettles, then of Columbia University College of Physicians and Surgeons, reported that with the aid of phase-contrast microscopy he could separate dried, unstained sperm into two distinct morphologic types. One type had small, round heads, and the other had larger, elongated heads. Shettles further claimed that he could count distinct chromosomes in some of the sperm heads, and that the smaller, round-headed sperm predominated in all specimens, though to different degrees. Shettles announced that the smaller sperm were the Y-bearers, and the larger-headed sperm the X-bearers.

Over the next half-year, Shettles' report was vigorously attacked by several well-known investigators in the field of human reproductive biology. Lord Rothschild, Dr. D. W. Bishop, and Dr. C. van Duijn, Jr., all expressed the firm opinion that Shettles' findings represented nothing more than optical artifacts, primarily caused by the fact that dried sperm on a slide are not nearly as flat as objects must be to permit proper visualization by phase-contrast microscopy. Another objection related to the fact that one of Shettles' photographed sperm heads measured thirty microns in length; the normal length is about five microns. In addition, Rothschild complained that Shettles had only estimated, rather than counted, the relative numbers of the two sperm types. Van Duijn remarked that no one had ever been able to visualize chromosomes in sperm with the use of either interference microscopy or electron microscopy, both more sensitive techniques than phase-contrast microscopy.

I would also add that it seems peculiar that Shettles insisted he could see only the two specific morphologic sperm types, with no intermediate forms. Most semenologists have reported finding innumerable variations in sperm head size and shape; some of these are illustrated in *Progress in Infertility,* by Behrman and Kistner (see references, chapter 6), on pages 618 and 619.

Of course, the upshot of the polemic was that neither side convinced the other. Despite the fact that to this day neither Dr. Shettles nor anyone else has been able to karyotype a sperm cell, Shettles still insists that his observed sperm-head dimorphism definitely reflects the crucial difference in sex chromosome constitution.

In the late 1960s, with the advent of fluorochrome staining of chromatin, it became apparent that this technique would soon be applied to spermatozoa. In 1970, the English doctors Peter Barlow and C. G. Vosa reported that Y-bodies were visible in about 45 percent of sperm stained with quinacrine mustard and viewed under fluorescent illumination (Fig. 7.1). The variance from the expected 50 percent should not be taken to represent a true deficiency of Y-bearing sperm, since technical problems in staining and viewing invariably cause the actual number of Y-bodies seen to be less than the theoretical level. That sperm with fluorescent bodies are indeed the Y-chromosome bearers was demonstrated by Drs. Sumner, Robinson, and Evans of Edinburgh. Using microdensitometric techniques, they established that sperm with Y-bodies had about 3 percent less DNA than those without. Staining for Y-bodies might be a useful technique to separate sperm, except for one problem: the staining process kills the sperm, rendering them useless for insemination.

Currently, two techniques for the separation of living sperm are receiving research attention. One of these is immunologic, involving

Fig. 7.1: Spermatozoan stained with quinacrine mustard. The Y-body is apparent. (Courtesy of Dr. Philip Fialkow and Mrs. Jean Bryant, University of Washington School of Medicine.)

an antigen-antibody reaction. In mice, H-Y is one of the protein antigen systems responsible for the recognition and rejection of "foreign" tissue. Since the gene for H-Y is located on the Y chromosome, only male mice possess these antigens. Therefore, Drs. D. Bennett and E. A. Boyse of New York caused female mice to form antibodies to H-Y by grafting skin from males to the females. Next, they bled the females to obtain blood serum containing anti-H-Y antibodies; then, they exposed mouse sperm to this serum. They reasoned that the antibodies should attack and destroy only Y-bearing sperm. In fact, after artificial insemination with the treated sperm samples, the percentage of males in the offspring was only forty-five, compared to the usual 53 percent in mice. This 8 percent difference is not spectacular, but it is significant, and may presage a usable technique to select for female offspring.

A method for the selection of male children might evolve from the work of Drs. Ericsson, Langevin, and Nishino of Berlin. These researchers layered a solution of sperm over a medium containing a quantity of the protein, albumin. They found that Y-bearing sperm were far more adept than the X's at crossing the interface between the sperm solution and the more dense, viscous albumin mixture. Furthermore, the difference could be intensified by running the separation process in successive steps, progressively increasing the concentration of albumin. After a three-step procedure, the Berlin group obtained a solution with 85 percent Y-bearing sperm, as shown by fluorescent Y-body analysis. Rabbit semen exposed to this separation procedure proved to be fully fertile.

Unfortunately, however, the future utility of the procedure is unsettled. In recent months, two highly regarded teams of geneticists have published reports stating that they were unable to confirm the results obtained by the Ericsson group; after passing semen samples through albumen layers, approximately equal numbers of X- and Y-bearing sperm were found. Further experiments are obviously in order.

Empirical Methods of Sex Determination

Although none of the techniques described above has yet proved to be of practical use, some reproductive biologists have attempted to apply some of the principles involved to the development of empirical clinical modalities for the preconceptual determination of fetal sex. In this area, no one has been more active than Dr. Landrum Shettles.

Dr. Shettles proposes the following prescription for couples who wish to conceive a boy: (1) Have intercourse near to the time of ovulation (as determined by daily recordings of body temperature over the course of a few months). In addition, there should be deep vaginal penetration at ejaculation. Shettles states that these two maneuvers will give the advantage to the small, round-headed sperm, which he believes to be the faster-swimming Y-bearers. (2) The woman should use a baking-soda douche before intercourse, and she should achieve orgasm. Shettles believes that survival and motility of "male" sperm are preferentially increased in an alkaline environment; hence the need for the baking soda. Orgasm is recommended because of the claim that alkaline secretions are released at this time. The Shettles formula for female children is the reverse of that for males: (1) intercourse two or three days prior to ovulation, with shallow penetration (to give the presumptive advantage to the slower-moving but longer-lived "female" sperm), and (2) a pre-coital acid douche of vinegar and water, and avoidance of orgasm by the woman. Dr. Shettles has publicized his methods in an article in *Look,* and in a book entitled *Your Baby's Sex — Now You Can Choose,* which he wrote with David Rorvik.

There are several reasons why one might be a bit wary about unqualifiedly accepting Dr. Shettles' suggestions. First, he has never provided proof that his reported sperm-head dimorphism does in fact reflect chromosomal dimorphism. His conclusion to this effect thus remains an unsubstantiated claim. Furthermore, Untermeyer's theory of the supposed effects on sperm of acid and alkali has been repeatedly questioned since its introduction more than forty years ago. For example, in 1971, Dr. R. B. Diasio and Dr. R. H. Glass used fluorescent Y-body staining to show that sperm migration is totally unaffected by the acidity of the solution in which the sperm happen to find themselves. In addition, intrinsic sperm motility probably plays a very minor role in the advance of the sperm toward the egg. For one thing, semen liquifies shortly after intercourse, so that even with shallow penetration, as the watery semen runs into the deepest recesses of the vagina, sperm would be indiscriminately provided a "free lift" toward the cervix. Also, the fact that sperm have been found in the human fallopian tube within thirty minutes of insemination speaks strongly against differential sperm motility as the important factor in sex determination: at the usual rate of intrinsic sperm migration, it would take these tiny swimmers closer to thirty days than thirty minutes to negotiate the distance from vagina to mid-fallopian tube.

With regard to Shettles' recommendations for the timing of intercourse, some interesting data have been presented by the Colombian Dr. Rodrigo Guerrero. Dr. Guerrero analyzed the time of insemination in 1,318 conception cycles in which body temperatures had been taken for either "rhythm" contraception or the treatment of infertility. From this information, he determined that when *artificial* insemination was employed very close to ovulation, 62 percent of the offspring were boys; however, when AI was performed three or more days prior to ovulation, only 39 percent boys resulted. These figures agreed closely with AI data previously reported by Dr. Sophia Kleegman. On the other hand, when *natural* insemination was used, the reverse situation was found: 44 percent boys with intercourse close to ovulation, 68 percent boys with intercourse several days before ovulation. This finding — exactly the opposite of Shettles' claim — held true in data from five different countries and five major cities in the United States. No specific explanation can yet be offered for this strange insemination-related difference.

A 1970 advertisement for Dr. Shettles' book makes mention of the "hundreds of happy parents" who had benefited from the "scientifically proven" procedures. However, in the same year, Shettles claimed to have personally used his technique only 41 times with 35 successes over a twelve-year period. It's difficult to understand why Dr. Shettles, after twelve years' work, was able to report on only forty-one couples. If I were to announce tomorrow that I had at my disposal a highly effective and safe technique for preconceptual sex determination, and wished to have volunteers to certify its utility, I might not survive the stampede at my office the next morning.

At present, a large number of reproductive biologists and obstetricians remain skeptical of Shettles' claims and recommendations, and do not advise their patients to utilize his sex-selection plans. Future medical historians will have to decide whether Shettles is a great unappreciated genius or a crackpot.

A German economics professor, Dr. Otfried Hatzold, is currently in the process of analyzing the applicability of a slightly different form of timed intercourse for sex determination. Instead of relying upon shifts in body temperature to signal ovulation, Hatzold is proceeding under the assumption that ovulation occurs fifteen days prior to the menstrual period. (This is generally, but not invariably, true.) Over a preliminary interval of a year, he instructs his patients to keep menstrual calendars, and then, by subtracting fifteen days from the first day of each menstrual period, he determines the earliest and the latest days on which ovulation has presumably occurred.

Next, proceeding from the Shettlesian premise that male sperm move faster but live a shorter period of time, Hatzold advises his couples who want a girl to have intercourse two days before the earliest possible date of ovulation. For a boy, the couple is told to have intercourse right on the day of the latest possible ovulation. Dr. Hatzold told me that eventually he intends to work with some two thousand couples, but that to date, he has too few cases to provide meaningful statistical data. It will be of interest to see his final figures. Until they are available, and until other workers have tried to corroborate his findings, no reasonable conclusions can be drawn.

Preconceptual Sex Determination:
Implications and Future Directions

Sociologists, psychologists, and demographers have indulged in considerable speculation regarding the potential problems and novel situations that might arise from the use of preconceptual sex determination. Before discussing these issues, I should mention two general considerations which are basic to all further thoughts. The first consideration is the effectiveness of any technique. A procedure with a success rate of 100 percent will have greater impact than one with, say 70 percent effectiveness. The second primary consideration is the number of persons who would avail themselves of sex-choice procedures, a number that would, no doubt, depend upon both the reliability and the nature of the techniques. For example, a do-it-yourself method to predetermine sex would certainly achieve wider use than one necessitating the intervention of a third party, such as an artificial inseminator.

I believe that any societal effects of preconceptual sex determination would be gradual, rather than sudden, since the earlier techniques are likely to be considerably less than 100 percent effective, and since human conservatism would likely play its customary part. In a large survey of married women, 47 percent stated that they would not use sex-choice procedures, even if such were available, and another 15 percent expressed uncertainty. Furthermore, at this time, about half of all pregnancies are unplanned, which would tend to further dilute any general effect of purposeful sex determination.

The major concerns that have been expressed over the use of sex determination relate to the possibility of significant alterations in the sex ratio. All studies clearly indicate that the result of universal sex-choice satisfaction would be the birth of a considerable excess of boys. In most foreign countries and in many American subcultures, the male excess would be overwhelming. Amitai Etzioni considers that this might result in a return to "frontier-like attitudes," with the larger number of males causing increased lawlessness, lowered morality, and less consumption of culture. It has also been suggested that such an altered sex ratio might produce increases in prostitution and male homosexuality. Severe disturbances of current marital norms have been prognosticated, including "raiding" of young women and girls by older men, and serial or simultaneous polyandry.

However, there is far from universal acceptance of these upsetting predictions. Sociologist Charles Westoff and demographer Ronald Rindfuss present the opinion that the introduction of a reasonably reliable technique for sex determination would lead to an initial 20 percent excess in male births. Nevertheless, they say, the aberration in sex ratio would be short-lived, and would be balanced out by a secondary excess of female births. Smaller oscillations would then follow, with eventual normalization of the sex ratio. Furthermore, many students of social structure have pointed out that societies possess considerable adaptability, and would not of necessity be destroyed by altered sex ratios. For example, the sex ratio in Alaska is 132:100, a sizable excess of males. Yet, we hear nothing about increased homosexuality on the ice floes; morality would appear to be no lower in Fairbanks than in Fort Wayne; and as for being mugged or murdered, I'd rather take my chances on the streets of Anchorage, than those of Atlanta.

Another group of potential problems has to do with the sex-chosen children. For one thing, most users of sex choice probably would select a boy first, followed by a girl. What would be the effect on younger sister, should she learn that she had been "second choice"? For another thing, procedures of less than 100 percent effectiveness might create difficulties. There are already numerous sad cases on record of the psychological destruction of children who were not of the desired sex. What would be the fate of a child whose parents actually had put in an order for a boy, but had gotten a girl? Or vice versa?

Some people have expressed concern about possible government intervention in sex choice. If the number of female births begins to fall off, would lobbyists for the cosmetics and ladies-garment industries put pressure on congressmen to restrict and regulate the use of techniques for sex determination?

All the hypothesized consequences of sex determination would not necessarily be inimical to human well-being. An interesting fact is that the sex ratio of last children is 117:100. In addition, parents whose first two children are of the same sex are more likely to have another child than those with one boy and one girl. Therefore, it is possible that free use of sex determination techniques would lead to smaller families and a lower birth rate. However, the pessimistic interpretation of these same data is that sex choice might raise the birth rate, since many couples may simply take their chances on the first two pregnancies, resorting to sex choice only if the first two children are of the same sex. This eventuality would increase the number of three-child families.

In summary, all this makes for interesting speculation, but it can be no more than that. Once again, we can see the futility of trying to cross unbuilt bridges. I think the most reasonable way to proceed is by the usual trial, error, and readjustment; the odds of doing ourselves irreparable harm in the process appear small indeed.

From the practical viewpoint, what of sex selection today and in the future? Basically, there are two major groups of persons who would want to use sex-determination techniques. The larger number of users are those interested in family planning. Admittedly, this is not a medical necessity, but I don't hold the belief that the gratification of "nonessential" desires is necessarily sinful. I think that a mechanism to permit people to have as many of the kind of children that they want, when they want them, would be a fine and welcome advance.

The second, and smaller, group of persons for whom sex-choice techniques might be helpful would be those at risk for the transmission of deleterious X-linked genes. Female carriers could avoid having affected sons by selecting for girls at conception, thereby precluding the necessity for amniocentesis and sex-specific abortion (chapter 5). On the other hand, affected men wishing to reproduce could select for boys, all of whom would be normal. By not having girls, all of whom would be carriers, a hemophiliac man could be certain that none of his descendants would have to worry about his "hemophiliac gene."

As previously mentioned, no tested, effective method now exists for the separation of living, undamaged ejaculated human sperm. Furthermore, I believe that Shettles' recipes are not firmly grounded scientifically and are far from proved on an empirical basis. Evaluation of Hatzold's data is not possible until his final results become available and are confirmed by other investigators.

However, there seems to be no reason why couples might not try the clinical methods, as long as they would be willing to accept the "wrong result." Certainly, a husband and wife with two boys who would want a third child *only* if it were a girl would be best advised to stand pat. But, there could be no real objection to the use of the clinical techniques by a couple determined to try a pregnancy in any case, but wanting to maximize chances of the desired type of child. This approach would seem especially reasonable for a couple where the wife carries an X-linked recessive gene, and who will, in any event, undergo amniocentesis. *It is important to emphasize that attempted preconceptual sex determination should not at this time be advocated as a substitute for amniocentesis for X-linked diseases.*

The specific clinical options currently open are the Shettles natural insemination modalities, and the timing of artificial insemination based on the figures of Kleegman and Guerrero. At this time, the AI techniques appear to have the stronger supporting data. Couples wishing to have boys might arrange for artificial insemination with the husband's sperm on the day of presumed ovulation. Those wanting girls could be inseminated three days in advance of ovulation.

Are the clinical techniques in fact perfectly safe? That, no one can say. While damage to the sperm, and therefore the child, seems unlikely, there exists no evidence to certify that acid or alkaline douches are in fact innocuous. In addition, the theory proposed by Dr. James German to explain the causation of Down's syndrome centers around the hypothesis that fertilization might occur long enough after ovulation for the egg to have suffered degenerative chromosomal damage. If this theory is in fact true, then trying for a boy by the clinical techniques (involving avoidance of insemination well in advance of ovulation) would increase the likelihood that the child might have Down's syndrome. But, much theoretical and some experimental evidence exists in opposition to German's theory.

A safe and reliable technique for preconceptual sex determination may be developed at any time, either as the result of refinements of present procedures, or as a sudden and unexpected breakthrough, perhaps a spin-off of research in some other aspect of reproductive biology. Until that time, I have to disagree with Shettles and Rorvik, and say that I think we're stuck with the "50-50 Club," or at best, with timed artificial insemination, perhaps a 60-40 Club. A good deal of careful basic and clinical research, followed by sober analysis of data, should precede any attempts to enroll people in a 100-0 Club.

8

Ectogenesis

Most people are suspicious of innovations and tend to resist them. This is especially true when the innovations involve emotionally charged aspects of human behavior. Hence, the tremendous negativism and opposition which continually surface to artificial insemination. However, the battles over the wisdom and propriety of AID are as nothing in comparison to those now being fought over ectogenesis.

Ectogenesis (ecto: outside; genesis: beginning) is a broad subject; it may be considered to include a variety of methods and techniques which permit an egg or an embryo to complete some or all of its prenatal development outside, rather than within, the female reproductive tract. For various reasons, the thought of an embryo's spending part or all of its antenatal existence outside its mother's body is highly upsetting to many people. During the past seven or eight years, ectogenesis has been precipitating hideous ructions, with a colorful assortment of savants desperately trying to outdo each other in the generation of hysteria in lecture halls, newspapers, and lay and professional journals. Columnist Nicholas von Hoffman believes that ectogenesis represents an attempt to adjust man to the needs of technology. Leon Kass reminds us of the Arab belief that a camel permitted to stick his nose into a tent will soon move his entire body inside. To Kass, AID represents the camel's nose, and ectogenesis the camel's neck; he fears that the next step may be the camel's spermatozoa. Paul Ramsey pontificates over the unspeakable horrors that he feels must inevitably attend the unnatural reproductive behavior of those who would play God. Throughout all the discussion is interlaced the specter of *Brave New World,* and even the ghost of Mary Shelley has been spotted here and there, flitting across the landscape.

Such attitudes are excessive and inappropriate. The social engineers have been building their bridges far in advance of the developments of the genetic and reproductive engineers. All the greatly feared potential consequences of fertilizing an egg in a plastic dish may very well never occur. The usual stepwise nature of medical and societal progress probably will prevail: preliminary experimentation will lead to clinical use on a limited scale, which, in turn will generate further data. By the time expanded clinical use comes under serious consideration, the public will have had time to assimilate the concepts involved, thereby dispelling the novelty and disposing of the initial horror. All things in good time.

In this chapter, we'll review the processes of natural ovulation, fertilization, and embryogenesis; then, we'll discuss various techniques of ectogenetic ovum maturation, fertilization, and embryo development. Lastly, we'll review some of the potential applications and the moral-ethical aspects of ectogenesis in the human.

The Natural Course of Ovulation, Fertilization, and Embryogenesis

At birth, the human female possesses in her ovaries all the eggs she will ever have. Throughout her reproductive years, the pituitary gland, located at the base of the brain, releases gonadotrophic hormones in a cyclic manner, which causes waves of ripening among the eggs. Each month, several eggs respond to the pituitary hormones, and approach the ovarian surface. Then, approximately halfway between two menstrual periods, the pituitary gland releases a burst of a substance called luteinizing hormone (LH). For some reason, only one of the preovulatory eggs usually responds to the LH by breaking free through the surface of the ovary (ovulation); from here the egg enters the fallopian tube. The other preovulatory eggs degenerate, and a whole new batch will be brought into readiness by the next month.

Approximately a day and a half passes between LH release and ovulation. During this time, some important changes occur in the chromosomes of the egg. Remember that eggs achieve a chromosome number of twenty-three by first replicating their chromosomal DNA, and then undergoing two cell divisions, a process known as meiosis (Introduction). DNA replication in eggs takes place during the female's own fetal life, and the chromosomes then remain in this duplicated condition until many years later, when the egg ripens and LH is released. LH causes meiosis to resume: the first division takes place, with half the chromosomal material being shunted into the tiny first polar body, which soon degenerates. Then, in the egg, the second meiotic division begins, but progresses only half-way to completion. Here, the process arrests again, pending fertilization.

Fertilization occurs at the junction of the upper and middle thirds of the fallopian tube. By as yet undetermined mechanisms, sperm are rapidly transported from the vagina, through the uterus, and into the tubes. Many sperm attach to the outer wall of the egg, but only the first one is able to penetrate inside. It is believed that the entry of the first spermatozoan triggers a chemical reaction at the egg surface, which repels all other potential invaders. This is necessary, because fertilization by more than one sperm cell would cause a lethal chromosomal imbalance in the resulting embryo.

Sperm entry also causes resumption of meiosis. The second meiotic division proceeds to completion, with extrusion of half the remaining DNA into the second polar body, leaving twenty-three chromosomes in the egg. This group of chromosomes gradually moves toward the group of twenty-three sperm chromosomes; finally the two groups merge, reconstituting the normal chromosome number of forty-six. The chromosomes then replicate their DNA, after which the one-cell embryo divides into two cells, each with forty-six chromosomes. These processes of fertilization and cleavage of the one-cell embryo are illustrated in Figure 7 of the Introduction.

After fertilization, the embryo remains in the fallopian tube for about three days, during which time its cells divide until there are about sixteen to thirty-two; this solid ball of cells is called a *morula* (Fig. 8.1a). During its residence in the tube, the embryo draws sustenance from the fluid secreted by the tubal lining. Upon entry into the uterus, the embryo subsists upon uterine fluid for another four days, and then implants in the uterine lining. Thus, for a week after fertilization, the embryo is a free-living structure, with no physical attachment to the mother. By the time of implantation, the embryo has divided into more than one hundred cells, which are arranged around a fluid-filled central cavity known as a *blastocele*. At this stage, the embryo is called a blastocyst (Fig. 8.1b), and its structure demonstrates the earliest manifestation of cell differentiation, or specialization. The cells making up the periphery of the blastocyst are *trophoblast:* these cells will invade the uterine wall and form the placenta. A relatively small cell group between the trophoblast and the blastocele is called the *inner cell mass.* It is these cells which will in time give rise to the embryo proper.

During the week that the embryo lives free in the tube and the uterus, the ovary causes some necessary groundwork to be laid down. After ovulation, the cells which formerly surrounded the egg *(follicle cells)* are transformed by LH into the *corpus luteum,* a structure capable of secreting the hormone progesterone. Progesterone reaches the uterine lining by way of the blood, and produces specific changes in the lining which cause it to be receptive to implantation by the embryo. Progesterone is necessary for both implantation and maintenance of

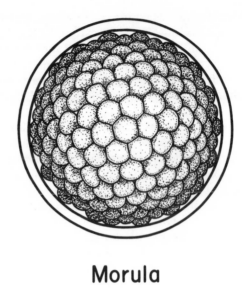

Morula

Fig. 8.1a: Appearance of mammalian embryo at morula stage — a solid ball of cells.

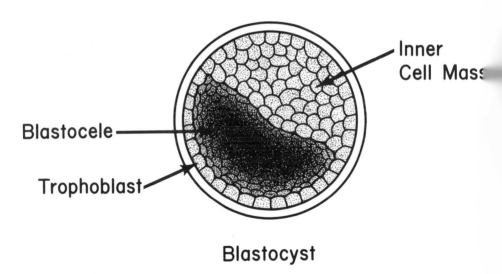

Blastocyst

Fig. 8.1b: A mammalian blastocyst. Note the hollow central cavity, the blastocele.

pregnancy. After implantation, the corpus luteum continues to produce progesterone for about six weeks, but after this time, sufficient progesterone is produced by the placenta itself, and the corpus luteum gradually involutes.

A week after fertilization, uterine implantation is initiated when the trophoblastic cells of the blastocyst make contact with the uterine lining, and aggressively invade this structure. Tapping into maternal vessels, they obtain nourishment from the mother's blood, and then send out lines of cells which anchor the embryo to the uterus. The trophoblast rapidly differentiates into the placenta, a specialized organ for extracting nutrients and oxygen from the uterine blood vessels, conveying them to the embryo, and returning to the uterus the embryo's wastes and excess carbon dioxide.

Now begins a most critical period for the embryo. During the next two to three months, the little inner cell mass proliferates and differentiates into the characteristic tissues and organs of the human body. By the end of the third month of pregnancy, all the organ systems of the fetus have formed, and for the remainder of the pregnancy, they will simply grow and mature. This explains why the first three months of pregnancy are the crucial ones in regard to birth defects. Drugs, infections, radiation — any harmful environmental agent — which acts on the embryo at this time may kill some of its cells, and thereby interfere with the normal formation of one or another organ, the result being an external or internal deformity. After completion of organ formation, however, cell death can no longer produce structural abnormalities. Before implantation, the effect of adverse environmental influences tends to be an all-or-nothing affair. If enough of the undifferentiated cells of the early embryo are killed, the embryo will die. On the other hand, if relatively few cells are killed, the survivors usually can regenerate sufficient numbers before the onset of differentiation.

This, then, is the natural course of affairs, Now, I will consider possible ectogenetic modifications. For ease of discussion, I'll divide the subject into three developmental phases: the period of ovum ripening and chromosomal maturation (meiosis); the period between fertilization and implantation; and the postimplantational period. However, it must be remembered that embryonic and fetal development is really a continuum, and that any combination of natural and ectogenetic maneuvers is theoretically possible.

Before going further, it would seem reasonable to define the frequently used term *in vitro*. Strictly speaking, it means "in glass," referring to laboratory vessels. "*In vitro* fertilization" means fertilization in a laboratory vessel (which, by the way, is more commonly plastic than glass). *In vitro* is often used in contradistinction to *in vivo*, which means "in the living (or natural) state." Thus, *in vivo* fertilization is that which occurs in the fallopian tube. For practical purposes, *in vitro* and ectogenetic can be used as interchangeable adjectives.

Ectogenesis Prior to Fertilization

Oocyte (or ovum) maturation is the term applied to the preovulatory meiotic process which is a necessary prerequisite to normal fertilization. As previously mentioned, natural meiosis is triggered by the same burst of pituitary LH that is responsible for ovulation. *In vitro* ovum maturation depends upon the fact that exposure to LH is not the only way in which maturation may be initiated. Much experimental work on eggs of rabbits and mice has shown that when an egg is surgically removed from its ovarian follicle, it will resume meiosis and undergo chromosomal maturation in a manner identical to that which occurs *in vivo* under the influence of LH. Moreover, the time sequences of *in vivo* and *in vitro* maturation seem to be similar. In mice, for example, chromosomal maturation is completed and ovulation occurs about fourteen hours after the pituitary LH stimulus; after removal of an egg from the follicle, about seventeen hours are required for the ovum to mature to the point where it may be fertilized.

For oocyte maturation to occur *in vitro*, the eggs cannot simply be left out in the open: they would quickly dry up and die. They must be maintained in special liquid culture media, much as is done with amniotic-fluid cells (chapter 5). Basically there are three varieties of culture medium: natural, synthetic, and combined. Natural media are unadulterated body fluids, such as blood serum or egg-follicle fluid. Synthetic media are those made up in the laboratory, where the investigators carefully control the concentrations of all constituents: water, sugars, proteins, fatty acids, inorganic ions. Combined media have both a synthetic and a natural component. The majority of media used today in ovum culture are synthetic or combined.

In ovum culture, as in any cell culture, composition of the medium is critical to success. With synthetic and combined media, depending upon the species whose ova are to be cultured, specific quantities of different substances are compounded. Concentrations of all components must be maintained within extremely narrow limits. An energy source is necessary: this may be a simple sugar or a short-chain metabolic acid such as pyruvic or lactic acid. Protein material is also important, and so are various inorganic ions (sodium, potassium, calcium, chloride, bicarbonate, etc.). Most modern combined media contain serum as the body-fluid constituent.

Composition of the culture medium is not the only consideration,

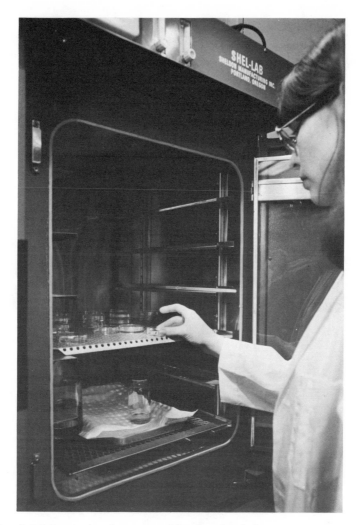

Fig. 8.2: A laboratory incubator, capable of maintaining constant environmental temperature, humidity, and gas concentrations, all necessary for successful embryo culture. (Photo by Jim Cummins.)

however. Strict control over other environmental factors is absolutely necessary. The temperature must remain constant, usually at about 37° to 37.5°C, which is normal body temperature. The concentrations of oxygen, carbon dioxide, and nitrogen in the air must be carefully monitored. The acidity of the medium is also critical: in most cases, optimal pH is about 7.3, slightly to the alkaline side of neutrality. To control all these environmental factors, use is often made of special incubators which can be set to provide proper temperature and

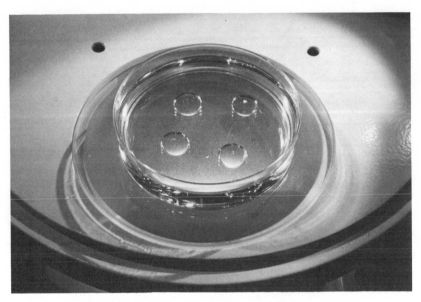

Fig. 8.3a: Plastic tissue culture dish, top view, showing droplets of culture medium under a layer of paraffin oil. (Photo by Jim Cummins.)

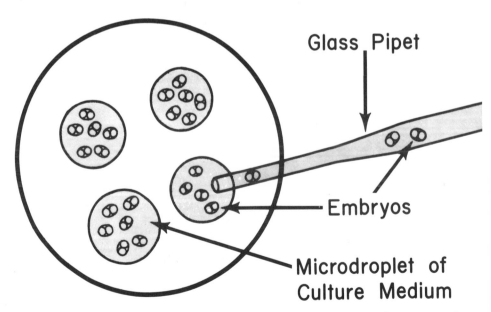

Fig. 8.3b: Drawing showing the manner in which eggs or embryos are manipulated within and between droplets, using glass pipet.

atmospheric gas concentrations (Fig. 8.2). The eggs themselves are placed in droplets of culture medium in plastic dishes, underneath a layer of inert paraffin oil; the latter prevents evaporation of the medium, and acts as a physical buffer, to avoid harm from inadvertent sudden changes in temperature or atmospheric composition (Figs 8.3a, 8.3b, 8.3c).

Finding a suitable culture system for oocytes (or any cells, for that matter) is largely a matter of repeated trial and error. This implies that even a culture medium which permits adequate cell growth, survival, and function can never be considered as the last word. A superior modification is always possible, and so, cell-culture specialists must be tinkerers by nature, constantly looking for the new sugar or protein which might be added to a culture medium to permit even better results.

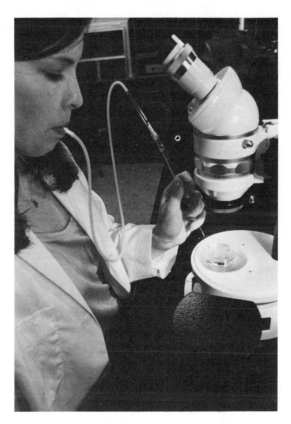

Fig. 8.3c: **Worker manipulating eggs and embryos in tissue culture dish, using dissecting microscope and finely-drawn glass pipet. (Photo by Jim Cummins.)**

By the early 1960s, ovum culture techniques had been extensively worked out in mice and rabbits. Then, Dr. Robert G. Edwards from Cambridge, while working at the Johns Hopkins Medical School, began to study human ovum maturation. Edwards removed eggs from surgically excised human ovaries, cultured them, and was able to report that a large number of the cultured eggs did indeed initiate maturation *in vitro*. He further demonstrated that the timing schedule was again consistent, full maturation being observed thirty-six to forty-three hours after liberation of the egg from the follicle. *In vivo*, human ovulation occurs about thirty-six hours after the LH burst.

Edwards' data were published in 1965. Three years later, Drs. Joseph Kennedy and Roger Donahue, also of Johns Hopkins, attempted to culture 426 human eggs, using several different media. Their results indicated that human oocytes are very difficult to culture. At best, only 33 percent of the eggs completed meiotic maturation.

Further attempts at *in vitro* ovum maturation have been made by Edwards and by several groups of American investigators. The results have been spotty: sometimes good, sometimes not. The consensus is that human ova are far more fastidious than mouse eggs in their *in vitro* growth and maturational requirements. Moreover, Edwards has established that oocytes matured *in vitro* give rise after fertilization to fewer viable embryos than eggs permitted to undergo maturation in the ovary, with recovery immediately prior to ovulation (method to be described in the next section). For this reason and because human eggs are not easy to come by in large numbers, Dr. Edwards and other workers in this field are currently concentrating their efforts on obtaining matured, preovulatory oocytes from the ovary. Then, they carry out fertilization and early embryonic development *in vitro*.

Ectogenesis from Fertilization to Implantation

Most people believe that *in vitro* manipulation of preimplantation embryos is of recent origin, but that's not the case. As long ago as 1913, Brachet studied cell division in five-day rabbit blastocyst embryos maintained in plasma. In 1929, Lewis and Gregory observed the development of early rabbit embryos in plasma, and a year later, Dr. Gregory Pincus, of birth-control pill fame, cultured rabbit embryos in combined media. In 1934, Pincus, with Dr. E. V. Enzmann, removed embryos from the fallopian tubes of rabbits, allowed them to divide and develop for a time *in vitro,* and then introduced them into the uteri of other rabbits, where some of them implanted and developed into normal progeny.

The recent major advances in *in vitro* fertilization and preimplantational ectogenesis have occurred as the result of the great strides made during the late 1940s and 1950s in tissue culture methodology. This accumulation of knowledge concerning culture media and techniques has enabled reproductive biologists to culture and manipulate eggs, sperm, and embryos with ever-increasing success.

At the present time, *in vitro* fertilization has been achieved in the rat, rabbit, guinea pig, hamster, gerbil, cat, squirrel monkey, mouse, and man. In addition, extensive experiments have been performed on the preimplantational embryos of rabbits and mice; much less work has been done on the preimplantational embryos of other species.

With the help of microscopes and tiny surgical instruments, embryos of laboratory animals have been subjected to various kinds of microsurgical procedures. Mouse embryos have been divided at the two-cell stage; upon intrauterine transfer and implantation, the two single cells have given rise to identical twins. Conversely, two one-cell embryos have been fused, the result being normal chimeric offspring with four genetic parents. (As shown in chapter 4, chimeras can also be produced by breaking up one embryo into its individual cells, and then injecting the cells into the blastocele cavity of a second embryo.) Sex diagnosis has been accomplished in preimplantational rabbit embryos by removing a few cells from the blastocysts, and analyzing these cells for female sex chromatin (Barr bodies). This method of sex diagnosis was quite accurate, as judged by the sexes of the embryos after implantation and development. Embryos of different species have been frozen, shipped half-way around the world, and then, when thawed, implanted in the uteri of recipients, later to be born normal.

The manipulation of preimplantational animal embryos reached its high point in 1972, when Dr. Anil Mukherjee reported his success in obtaining normal live baby mice after *in vitro* ovum maturation, fertilization, and development to the blastocyst stage. This was the first time anyone had been able to obtain live young after total *in vitro* preimplantational development. Mukherjee removed nonmatured eggs from the ovaries of adult mice, and cultured them through maturation to the point where they could be fertilized. Then, he transferred the eggs to fresh droplets of culture medium containing mouse spermatozoa. Some of the eggs were fertilized, and four days later, the resulting blastocysts were transferred to the hormonally prepared uteri of recipient female mice. The normal duration of pregnancy in the mouse is twenty-one days. Seventeen days after transfer of the four-day embryos, Mukherjee observed the birth of normal baby mice. This was indeed a striking achievement, but lest it be thought that it signaled the arrival of the millenium in preimplantational ectogenesis, it must be pointed out that of 325 ova initially subjected to culture, only 140 completed chromosomal maturation. Still more eggs failed to be fertilized, to develop into blastocysts, or to implant. In the end, only five newborn mice were obtained, representing a yield of 1.5 percent. Technically, there is room for considerable improvement.

Since human eggs are available in smaller numbers and are more fastidious in their culture requirements than the oocytes of laboratory animals, the achievements to date in human preimplantational ectogenesis are less spectacular. Yet, they have managed to attract widespread lay and professional attention. Dr. Robert G. Edwards and his gynecologist-collaborator Dr. Patrick Steptoe, both of England, have performed by far the greatest amount of work in this field, and my discussion will focus largely upon their procedures.

Edwards and Steptoe have concentrated their efforts on couples who are infertile because of intractable blockage of the wife's fallopian tubes. The ovaries in these women produce normal eggs, which are normally ovulated, with formation of normal corpora lutea. The corpora lutea secrete progesterone, and thereby cause the uterine lining to be normally receptive to implantation. The only problem is the tubal obstruction, usually a sequel to a previous episode of pelvic infection. This blockage prevents the sperm from reaching the egg, and the egg from reaching the uterus.

Therefore, Edwards and Steptoe have been trying to use ectogenesis to circumvent the obstructed tubes. Because of the difficulty in obtaining human oocytes, and the relatively poor results of human *in vitro* egg maturation, the English team has chosen to work with eggs matured *in vivo*. First, the woman is given a series of injections of Pergonal, the so-called "fertility drug." Pergonal is a mixture of the pituitary gonadotrophic hormones which cause eggs to ripen in the ovaries. After the Pergonal has acted, an injection is given of a substance called human chorionic gonadotrophin (HCG). HCG is a hormone isolated from human placental tissue; structurally, it is much like LH. Therefore, upon injection, it can trigger chromosomal maturation and ovulation.

Ovulation, however, is not permitted to occur. Approximately thirty-two hours after injection of HCG, the woman receives general anesthesia, and Dr. Steptoe goes to work with an instrument called a laparoscope. This is a long, thin metal tube, encasing a bundle of illuminated optical fibers. The far end of the tube is inserted into the patient's abdomen, through a small incision below the navel. Then, using a lens at the near end, the operator can inspect all the internal organs. The laparoscope enjoys wide use as a diagnostic tool in situations such as unexplained abdominal or pelvic pain. It is also used to accomplish the so-called "band-aid" tubal sterilizations. Dr. Steptoe, however, turns his attention to the ovaries, and looks for the little blisterlike swellings on the surface, which represent sites from which ovulation may occur four to six hours later. On the average, one potential ovulation point is seen per ovary. Upon finding one of these points, Dr. Steptoe keeps the spot under vision, and inserts into the abdomen a specially designed thin suction apparatus. With this, he punctures the swelling and draws out the ovum in its follicular fluid. The ova so collected are given to Dr. Edwards, after which the procedure is terminated, and the woman brought out of anesthesia. Usually, there is little postlaparoscopy discomfort.

The preovulatory eggs received by Dr. Edwards have almost, but not quite, completed chromosomal maturation. They are carefully put into culture medium in an incubator, and the fertilization culture is prepared. Experience has shown that the fertilization medium should be a little more alkaline than medium for the culture of eggs or embryos. The husband provides Dr. Edwards with a sperm sample by masturbation. The spermatozoa are washed free of the seminal fluid, and are then diluted in culture medium to a concentration of about one million per milliliter, which seems to be optimal for fertilization. After four hours of oocyte culture (thirty-six hours after HCG; the presumed time at which ovulation would have occurred), the eggs are placed into the sperm suspension. Sperm penetration usually occurs within three or four hours, but the fertilization process is allowed to continue for about twelve hours. At this time, the fertilized, one-cell embryos are taken out of the fertilization medium, and transferred into the medium which Dr. Edwards has found to be most suitable for culture of the developing embryo. Here they remain, under strictly controlled environmental conditions, for about three days.

During the time that the embryos are being cultured to the blastocyst stage, Dr. Edwards' team gives the wife injections of progesterone, in an attempt to supplement the secretions of the new corpus luteum. Edwards and Steptoe hope that these injections will increase the likelihood that sufficient progesterone will reach the uterine lining, causing it to be maximally receptive to implantation by the embryo.

Edwards claims to have successfully fertilized more than seventy human ova. How many of these were cultured to the blastocyst stage is not known, but Edwards has published photographs of two apparently normal blastocysts. He and Steptoe have attempted to transfer at least ten blastocysts back into the women from whom the eggs were obtained. The transfer procedure consisted of inserting a fine plastic catheter through the cervix and into the uterus; then, the embryo was injected through the catheter. Although in some cases the menstrual period was delayed by a few days, none of the embryos took hold and developed, within the uterus; however, a recent publication describes a single instance where the embryo seems to have grown for a time in the wall of the fallopian tube.

So, for all their experience and knowledgeability, Edwards and Steptoe have yet to experience a clinical success. Has anyone else? Right now, no one can be certain.

Dr. Landrum Shettles has been involved in this business, too. He claims to have fertilized a human egg *in vitro*, and then to have cultivated the embryo to the blastocyst stage, using mucus from the cervix as culture medium. Shettles has declared his intention to implant an embryo, but apparently has not yet succeeded in doing so.

In 1973, a group of Australian doctors headed by Dr. D. DeKretzer reported the transfer of an embryo at the eight-cell stage. The ovum had been taken from the ovary shortly before ovulation, and then had been fertilized and cultivated *in vitro*, according to the Edwards-Steptoe formula. Transfer was performed through the cervix, seventy-four hours after fertilization. Although urinary hormone excretion levels suggested that the embryo was functioning, this could not be ascertained. In any case, menstruation occurred nine days after transfer.

Scientific accomplishments are customarily announced in professional journals or at research meetings. Therefore, Dr. Douglas Bevis, of Leeds, England, stirred up quite a controversy by mentioning to a group of reporters that he knew of three normal babies who had been born after *in vitro* fertilization. Bevis' claim came during the meeting of the British Medical Association, in July, 1974. Despite the fact that Dr. Bevis is an eminent physician, rightly renowned for his pioneering research in the field of Rh disease, he was universally condemned by the scientific community because of the manner in which he presented his information. Reproductive biologists around the world — including, of course, Drs. Steptoe and Edwards — demanded to see scientific proof of his claim.

Although such evidence may now be in print, at this writing, no proof has reached the world. Furthermore, when he made his announcement and shortly thereafter, Dr. Bevis told slightly different versions of his story. At first, he claimed not to have been a member of the successful team; later, he stated that he alone had done the work, that it had taken thirty-six attempted transfers to achieve the three successes, that he had used a technique similar to Edwards' and Steptoe's, and that he had chosen to announce his results in that unorthodox fashion in order to give privacy to the "research team" and the babies. This last statement was ludicrous; in the few days following his announcement, Bevis was so beseiged by telephone pests that he announced in disgust that he would resign from further research work in infertility.

Thus, in the absence of a formal scientific report, we have no way to evaluate the truth of Bevis' claim. As it stands, the story has to be counted as one of the many peculiar and puzzling items of medical history.

Ectogenesis from Implantation to Birth

In contrast to the great progress made during the past decade in experimental preimplantational ectogenesis in humans and other animals, the current status of implantational and postimplantational ectogenesis is as a desert. Little work is now being done in this area, and what has been accomplished holds small promise in the way of clinical application during the near future.

From a functional standpoint, this research can be divided into two groups: those experiments leading to the development of an artificial placenta, and those designed to work out an artificial uterus into which an embryo might implant, using its own placental system.

The Artificial Placenta

A large amount of research on the cardiac, respiratory, and metabolic functioning of human fetuses obtained at therapeutic abortion led to the landmark paper of Drs. B. Westin, R. Nyberg, and G. Enhorning, of Stockholm. This work, published in 1958 in *Acta Paediatrica*, was entitled "A Technique for Perfusion of the Previable Human Fetus." Basically, the authors described their attempts to adapt the surgical heart-lung machine principle to the problem of sustaining life in fetuses.

After removal of a fetus from the uterus, the umbilical cord was cut, and plastic catheters were inserted into the severed ends of the umbilical blood vessels. In this way, blood from the perfusing system could be pumped to the fetus via the umbilical vein; at the same time, the fetal circulation would pump blood back to the perfusor through the umbilical arteries. Thus, the circulatory relationship was the same as normally obtains between fetus and placenta.

Next, the fetus was placed into a glass container filled with a glucose-water solution. This container was maintained at body temperature, and at strictly controlled pressure. This arrangement simulated the amniotic sac.

The perfusion apparatus itself consisted of a rotating-spiral oxygenator which served two functions: to circulate blood under proper pressure to and from the fetus, and to constantly replenish oxygen while withdrawing carbon dioxide from the blood, all the while maintaining the liquid at proper body temperature.

The Scandinavian team used their machine on seven fetuses weighing between four and twelve ounces, and were able to maintain life for up to twelve hours. This was a start, and it generated a good deal of enthusiasm in the pediatric world. By the mid and late sixties, many reports could be found in the medical literature, describing attempts to improve on the record of the Swedish doctors. Some experiments were performed on human fetuses, others on fetal lambs (which are good experimental models for human fetuses), and a few on rabbits and other animals. The oxygenators were modified, attempts were made to compose better synthetic amniotic fluids, and the fetuses were provided nutrients in addition to oxygen. However, striking success was never achieved in either human or sheep fetuses. At the longest, the fetuses could be kept alive for a day or two, but then they succumbed either to sudden cardiac arrest with circulatory collapse, or to imbalances in body metabolism. Both causes of death probably resulted from inability to properly clear the fetuses of waste materials. In other words, nature's placenta is still much more efficient than man's. What's more, this situation is not likely to change until after the performance of a good deal of basic research on placentas. This research will allow us to better understand the complex, precise, and adjustable functions of the placenta as fetal oxygenator, waste remover, and food supplier.

The Artificial Uterus

Since the present assortment of artificial placentas are so unsatisfactory, might it help to take the process of implantation a step back: that is, to design an extracorporeal site upon which a placenta could implant and then nurture its fetus in the customary fashion?

One theoretical possibility along these lines is the use of organ culture, to maintain a perfused mammalian uterus in culture medium under controlled environmental conditions. Unfortunately, it is proving no easier to perfuse organs than to perfuse fetuses, and survival currently is possible only in terms of a few days. In addition to difficulty in controlling an organ's metabolism, the culture systems have a strong tendency to become infected, despite all precautions to the contrary.

Another possibility is to discover a substance, which, when spread over the surface of a glass or plastic culture vessel, would permit implantation and growth of an embryo. This line of thought is illustrated by the work of Dr. Yu-Chih Hsu, of Johns Hopkins. Dr. Hsu coated petri dishes with connective tissue obtained from rat tails. Then, he was able to induce mouse blastocysts to implant in the connective tissue. After implantation, with assiduous daily adjustment of the contents of the culture medium, some of the embryos developed in more or less normal fashion to the mid-pregnancy stage. (Remember, mouse pregnancies last twenty-one days.) In a few cases, an apparently beating heart could be seen. Although scientists had for many years been culturing embryos for short periods on clotted plasma, connective tissues, and other substances, Hsu's experiment represented the first instance of successful implantation and subsequent postimplantational development.

However, this feat represents only a beginning. It will take tremendous amounts of painstaking labor to devise a system in which mammalian embryos can implant and develop to birth. In addition, normality of the offspring will have to be thoroughly assessed. Still, I believe that all this will be accomplished before anyone manages to come up with a satisfactory artificial placenta.

Potential Applications of Ectogenesis

Any discussion of the potential uses of ectogenetic research is likely to soon be outdated. Each new discovery opens the gates to new applications, and it's entirely possible that the most significant practical utilization of ectogenesis has yet to enter the mind of man.

However, the point should be made that during the past couple of years, there has been a noticeable decline in the number of scientific publications having to do with ectogenesis. This is partly due to consolidation and assimilation of the extremely rapid advances in the field from 1965 to 1972, and partly due to hostile social attitudes toward this research, which, in turn, lead to relative unavailability of funding for ectogenetic experimentation.

Potential Applications of Preimplantational Ectogenesis

As stated by Edwards and Steptoe, the most obvious use for preimplantational ectogenesis involves the treatment of infertile women. It has been estimated that more than 500,000 women in the United States alone may be infertile due to blockage of the fallopian tubes, and therefore theoretically eligible for the Edwards-Steptoe procedure of laparoscopic recovery of a matured ovum, *in vitro* fertilization, and intrauterine transfer. Although the procedure is being tried (and may have already succeeded) in humans, it should not yet be thought of as anything other than experimental. For one thing, the degree of safety has not been established in animals or in humans. For another, the success rate indicates that the technical aspects are anything but thoroughly established. Mukherjee's 1.5 percent incidence of live births in mice is very low, and, in addition, Bevis' unofficial claim of three successes in thirty-six attempted transfers (8 percent) is not only low, but represents a maximal estimation. Presumably, it does not include a number of women whose eggs could not be fertilized *in vitro,* or from whom it was impossible to recover eggs from the ovary.

Another practical negative is the fact that although almost any gynecologist can perform laparoscopy, it is not likely that the facilities and personnel necessary for *in vitro* embryogenesis will ever become widely available. Therefore, the capacity to perform the procedures probably will remain limited to major medical centers, with their attendant teams of clinical and laboratory specialists. Needless to say, it is also likely to be costly.

There are two variations of ectogenesis which deserve mention. One is the use of a donor egg, an approach that might be employed to permit a woman with undeveloped or surgically removed ovaries to have children. The other variation involves the woman who wants a child, but who, for whatever reason (illness, career obligations, vanity) does not wish to carry the pregnancy. Here, an egg from the woman, fertilized by the sperm of her husband, might be implanted in the uterus of another woman, who, for love or money, would go through the pregnancy and bear the child.

In either of these two situations, the problem is complicated by the fact that the uterine lining must be at just the right stage of progesterone-induced development to permit the embryo to implant. When the ovum donor and embryo recipient are different persons, there probably will be no properly timed corpus luteum to prepare the recipient's uterus. Consequently, there would have to be coincident induction of ovulation in donor and recipient, with care being taken that the eggs of the recipient are not fertilized. This arrangement would be necessary because no one has the least idea of the precise progesterone injection schedule that could satisfactorily substitute for the corpus luteum.

Preimplantational embryo culture might allow for sex determination by microbiopsy of a few trophoblast cells, with subsequent staining for X- and/or Y-bodies. This has already been achieved in the rabbit. Such a procedure would be of use to carrier women and affected men with X-linked diseases, such as hemophilia or muscular dystrophy, who wish to reproduce. Later, if and when the technique has been well worked out, it could even be utilized for straightforward family planning.

As mentioned in chapter 4, preimplantational ectogenesis might be of use in gene therapy. This could be done via gene transduction into the cell(s) of an early embryo, or through the creation of chimeras, either by embryo fusion or by injection of cells into a blastocyst cavity.

Eventually, if ectogenetic procedures prove safe and if techniques are developed for preimplantational screening for birth defects (such as micro-biopsies for chromosomal and/or biochemical surveys), preimplantational ectogenesis could evolve into the method of choice for the initiation of reproduction. Once people have become accustomed to the general idea, they might prefer to request fertilization and preimplantational culture *in vitro,* with subsequent implantation of a normal embryo. This would be a reasonable way to prevent many serious birth defects.

It must not be thought, however, that the only potential applications of ectogenesis are directly clinical. Even if no human baby ever results from *in vitro* fertilization, ectogenetic research has tremendous potential for bettering the human condition. Chromosomal diseases such as mongolism arise during ovum maturation or at fertilization: therefore, the culture of eggs and embryos would offer the opportunity to directly study the origins of these chromosomal errors. In addition, study of the processes involved in fertilization and preimplantational embryonic development might provide clues to the reasons for the occurrence of some birth defects and some forms of miscarriage and "unexplained" infertility. On the other side of the coin, such studies might lead to the development of better and safer contraceptives.

Experimental ectogenesis might also be of great value in the field of cancer research. In their morphologic appearance and their growth characteristics, cancer cells are more like embryonic than adult cells. In addition, some cancer cells produce characteristically fetal substances. For example, the same alpha-fetoprotein which is used in the prenatal diagnosis of anencephaly/spina bifida has been found to be produced by liver cancers. Basically, then, cancer cells seem to have dedifferentiated into a more primitive state. Conjoint studies of cancer cells and embryos might well lead to understanding of the basic nature of the de-differentiation process, which, in turn, could produce agents capable of curing cancer by reversing the de-differentiation. In addition, there might be recognized more compounds like alphafetoprotein, which would be of use in screening programs for the early detection of cancer.

Potential Applications of Postimplantational Ectogenesis

Ex utero development at this stage would also offer considerable research potential. Since the first six postimplantational weeks represent the time during which organ differentiation occurs, normal and abnormal morphogenesis could be studied. This might give us a good deal of information about the origins of many common birth defects. Moreover, study of implantation *in vitro* might help us learn more about miscarriage, a problem which occurs in approximately 15 percent of all recognized pregnancies. Many questions about pregnancy wastage might be answered: for example, what happens to all those "extra" male conceptions?

Prematurity is a major obstetrical problem. "Glass wombs," as Dr. Fletcher categorically puts it, might certainly prove to be most helpful both in caring for and in studying the problems of early premies, whose grossly immature lungs now condemn them to a rapid death.

Total *in vitro* postimplantational growth would eliminate the maternal morbidity and mortality currently attendant upon pregnancy. Birth trauma to fetus and mother would also be eliminated. Perhaps the incidence of birth defects could be reduced by ectogenesis, since fetuses would be less likely to be exposed to environmental *teratogens:* drugs, radiation, and viruses. Therefore, looking far into the future, if total prenatal growth *in vitro* should indeed prove safer than the natural method, it might well become the standard and preferred arrangement, at least for some people. Dr. Fletcher has said that those who love children and want to do right by them will pray for glass wombs. Though I can't quite share his evangelical fervor, I do think it would be nice to be able to exercise such an option. Perhaps our descendants will enjoy this advantage.

Moral and Ethical Issues Concerning Ectogenesis

The primary issue with regard to ectogenesis is that of the safety of the procedures utilized. No birth defects have been recognized in the small number of animals born to date after *in vitro* fertilization and/or preimplantational ectogenesis. Nor would such really be expected after manipulation of such early, undifferentiated embryos. Postimplantational ectogenesis, however, might carry a greater risk, since it is then that organ differentiation occurs, and a suboptimal environment could interfere with these processes, producing anomalies. In the case of both pre- and postimplantational ectogenesis, it would seem that considerably more work should be done in animals, standardizing the techniques and evaluating the offspring. Only then do I think the work should be tried in humans.

This last step, however, is one that Leon Kass feels should never be taken. His objection is that since the final test of safety must involve potential humans, it would constitute unethical experimentation upon the fetus and so could not be justified. He offers the opinion that ectogenesis cannot be considered therapeutic for the fetus, and analogizes ectogenesis to procedures which would involve deliberate attempts to produce damaged fetuses, such as artificial insemination with irradiated sperm, or the administration of thalidomide to pregnant women.

Kass, then, takes the position that it is preferable to have no life at all than to run the unknown (and possibly nonexistent) risks of ectogenesis. This position is challenged by Marc Lappé, of the Hastings Institute. Lappé sees the fetal risks of ectogenesis as less threatening than reproduction by a woman with PKU, or by two carriers for sickle cell anemia. Yet, no effort is made to forbid such persons to have children. Lappé's rebuttal need not be limited to the high-risk fetus, but could be extended to the general situation in humans. A 3 to 5 percent risk of serious physical or mental impairment is no small threat to a fetus, but few prospective parents give serious thought to this eventuality when deciding whether to have children.

In summary, I believe that human preimplantational ectogenesis should be discouraged until reassuring animal data have been collected. At that point, I'd see nothing wrong with offering the procedures to infertile couples, after thorough explanation, and with informed consent. If the offspring of these couples turn out to be as healthy as or healthier than naturally conceived babies, then the techniques could be offered on a wider scale. As for the question of giving consent for the unconceived fetus, I think if prospective parents can give consent for natural conception, there's no reason why they can't give it for artificial conception. With regard to postimplantational ectogenesis, there is a gigantic quantity of experimentation to be done before we can even begin to think of using the techniques in humans. But the procedural principles would seem identical.

Questions have arisen concerning the possible psychological effects of ectogenesis on the children and the parents. No doubt the first patients will be subjected to celebrity status, but this does not appear to be a fatal disease. Some people even seem to thrive on it. And if the procedures should become routine, then this "problem" would rapidly vanish.

Would parents accept an ectogenetic child as readily as a naturally grown one? No one can say with any certainty. But adoptive parents seem to be at no disadvantage in this respect, and they have far less prenatal interaction with their future child than would be possible with ectogenesis.

What of eggs or embryos which are not used for implantation into the donor? The questions have been raised as to whom they would belong and what uses might properly be made of them. However, to me, this issue seems straightforward. Like sperm samples, the eggs and embryos would seem to rightfully belong to the donor, who, upon donation, should specify how they may be utilized. For example, if three eggs were taken from a woman's ovary by laparoscopy, and one was implanted after fertilization, the woman might give permission for the extra embryos to be implanted into other women with no eggs of their own. Or, she might allow them to be used for research. On the other hand, she could request that they all be given the opportunity to implant in her own uterus. She could also insist that any embryos past a given number must simply be washed down the sink.

Some people find the prospect of this latter procedure disturbing. They consider it to constitute the taking of human life, no different from murder. I would answer that sperm and eggs are also human and alive, and, for that matter, so are the body cells grown in culture for various reasons (such as amniocentesis). Incidentally, these cells also have reproductive potential, as we shall see in chapter 10. Altogether, this is not an issue that can be resolved on rational grounds. Flushing away a human blastocyst would not be a compunctious act for me, but anyone who is concerned that this microscopic ball of cells might have a soul ought not to engage in ectogenetic work. It seems as simple as that.

In these days of concern over the population explosion, some say that it is not right to devise new techniques that may help sterile couples to have babies. This callous statement is typical of many of our current crop of would-be social directors, who profess to harbor limitless quantities of concern for the human race, but who obviously could not care less for the feelings of individual members of that body. The famous Indian gynecologist Dr. V. N. Shirodkar was once asked how he could justify his development of a surgical procedure whose sole application was to prevent a certain form of repetitive late miscarriage. Shirodkar replied that although it was true that most Indians had too many babies, there were some who did not have enough, and that it was his wish to try to help those with either problem.

Leon Kass has declared that improved methods of tubal reconstructive surgery would render ectogenesis unnecessary. This statement reveals tremendous optimism or tremendous ignorance. If Dr. Kass himself had ever tried to repair these delicate structures, he would not have said so assuredly that results are "bound to improve with practice." Even when the surgeon can manage to restore patency to the tubes, he cannot assure fertility: the functional capacity of the tubes to nourish an embryo may have been destroyed forever. Incidentally, the major surgery necessary to tubal reconstruction is not without hazard to the woman; furthermore, women with reconstructed tubes are at significantly higher risk for dangerous ectopic (tubal) pregnancies.

One problem that will have to be met is the issue of viability: the time at which the fetus can survive independent of its mother. With the advent of complete ectogenesis, will we decide that every embryo is viable, and must be saved? Or, will we consider viability to represent the capacity for life independent of all or any means of support? Working up by stages to complete ectogenesis should give us the time and perspective necessary to gradually adjust our attitudes and come to a consensus. Nevertheless, I suspect that there will be a grand and entertaining donnybrook. I regret that I probably will not be around to witness it.

The issue of surrogate uteri, or, as it has been put, Rent-A-Wombs, is a hot one. Some people fear that it may lead to exploitation of poor women; however, attempts to forbid the practice could just as easily prevent many women from obtaining what for them would be enjoyable and remunerative work. Several women have told me that they never feel better than when they are pregnant, and would arrange to be always in that condition if only they would not have to keep the babies! And as for risks and hazards, I'll only begin to listen to that argument if we forbid men to enter boxing rings or biologists to work with pathogenic bacteria.

Surrogate uteri probably will not be with us before embryo transfer into the uterus of the egg donor has been shown to be safe and reliable. If and when this occurs, some questions certainly will have to be answered. For example, it has been asked who will be the mother: the egg donor or the recipient? I'd think a parallel could be drawn to adoption, but difficulty might well arise if the recipient were to plead inability to complete her share of the bargain, and request abortion. Or, on the contrary, she might apply for custody of the child. Furthermore, what should be done if she were to decide to solve such a disagreement by running away and concealing her whereabouts? Would fetus-napping have to be defined into the criminal code? Then, at the other end, suppose the fetus should be born defective, and the genetic parents refuse to accept the child? Would they be obliged to assume legal custody? Logically, they should, but legally, who knows? The ubiquitous Dr. Kass tries to solve these futuristic Gordian knots through his customary fashion of employing a rigid moralistic system. Kass dismisses the entire issue as simply a new form of prostitution, an amazing approach even more outlandish than considering AID to be adulterous. Like the viability issue, however, I suspect the problems of the surrogate uterus will yield to the future, a bit at a time.

Ectogenesis frequently inspires people to claim in hysterical tones that *Brave New World* is coming, complete with governmental control of reproduction. Nicholas von Hoffman's previously cited newspaper column (chapter 2) exemplifies this attitude; so does an article by M. Stanton Evans in the magazine *Private Practice*. Evans' paralogisms leap directly from the relief of infertility to government-ordered reproduction; from the abortion of mongoloid fetuses to the directed breeding of supermen. He raves over the "license to kill," as exemplified by euthanasia and "permissive" abortion. His flight from rationality is made complete, however, when he comments on "the obvious linkup between the test-tube baby syndrome and the continuing effort to submit the medical profession to government regulation."[1] From there, it's a small jump to a tirade against the evils of governmental regulation of medical practice, and an explanation of how Hitler would have loved *in vitro* fertilization.

But this is ridiculous. For one thing, ectogenesis offers no advantages over conventional reproductive techniques in the (weakly grounded) concept of breeding supermen. For another thing, ectogenesis cannot be used to evil ends as long as reproduction itself remains a totally individual prerogative. And a government that can dictate our reproductive practices will be capable of inflicting far worse indignities upon us. Ectogenesis cannot lead to slavery; slavery must pre-exist before any misuse of ectogenesis can occur.

So, with all due respect for Dr. Kass' camel, I believe that ectogenesis, taken a step at a time, and allowing for adjustment of individual and societal attitudes, is capable of leading to much good, both in research and clinical situations. We shouldn't expect to solve all the theoretical problems in advance, nor should we try. As with any venture, it would seem more reasonable to go a short distance down the road, see how things work out, and then proceed from that point.

I'll close this chapter by engaging in one last joust with Dr. Kass. In his articles, Kass comments that ectogenetic reproduction is dehumanizing. By way of an answer, I'd like to quote a not-so-funny paragraph from the work of humorist H. Allen Smith. The following is from Smith's book about Tahiti, *Two-thirds of a Coconut Tree:*

> We had Christmas dinner with Ralph Varady and, knowing that I am going to Bora Bora tomorrow, he told me some of the romantic traditions of that tropic isle. During the war more than twenty thousand American troops were stationed there and many more thousands passed through. After they had gone home a census was taken and there were 126 babies whose fathers were American. Forty percent of these died before reaching their teens. They had been started off in life with good milk and wholesome bread and ample vitamins and other salubrious things from the PX. After the troops left these children were switched onto the traditional diet of the natives, including much raw fish and breadfruit. As a consequence many of them sickened and died, and it is unlikely that their fathers ever heard about it. Or cared.[2]

Even though these babies were conceived "naturally," does this little history exemplify properly "human" reproduction? Does it show careful planning, with due respect for fetal and maternal well-being and a good outcome for the child? Somehow I just can't seem to work up much indignation over the prospect of a few couples trying to relieve their infertility by having their gametes make union in a plastic tube instead of a fallopian tube.

9

Parthenogenesis

Parthenogenesis is development of an unfertilized egg into an embryo, fetus, or individual. The word is derived from the Greek *parthenos* (virgin) and *genesis* (creation); thus it may be taken literally to mean "virgin birth." This semantic confusion of sexual intercourse with fertilization sometimes leads to unjustified conclusions: for example, upon first hearing of parthenogenesis, many people believe that they have discovered a "scientific explanation" for the "virgin birth" of Christ. However, a little reflection on the mechanism of sex determination in humans leads to the realization that in the absence of fertilization, an embryo cannot have a Y chromosome, and so any human *parthenogenones* would have to be female. So much for that attempt to reconcile science and religion.

Parthenogenesis in Invertebrates and Laboratory Mammals

Experimental Parthenogenesis

The eggs of the sea urchin, which normally are fertilized and undergo embryonic development in sea water, have long been favorites of experimental biologists. In 1896, the German scientist Oskar Hertwig added strychnine to a solution of sea water containing unfertilized sea urchin eggs. This procedure caused some of the eggs to become *activated*, which means they initiated embryonic development. Hertwig's line of experimentation was continued and extended to amphibians by Jacques Loeb, who published his results in 1913 and 1916. Another noted student of the parthenogenetic embryology of marine invertebrates and amphibians was Dr. F. R. Lillie, whose book on that and related subjects appeared in 1923.

Parthenogenesis came into prominence in the 1930s with the work of Dr. Gregory Pincus. Pincus wanted to study the cellular changes which occur in an egg upon fertilization, and he thought that if Loeb's work could be extended to mammals, it might provide him with a useful tool for his research. He decided to attempt parthenogenetic activation and development of rabbit eggs.

With respect to ovulation, rabbits are significantly different from the majority of mammals, including man. Most mammals maintain a more or less regular ovulatory cycle: mice release eggs every fourth or fifth day, women about once a month. Therefore, fertilization is dependent upon properly timed insemination. Rabbits, however, do not have spontaneous ovulatory cycles. Their famous fecundity is largely due to the fact that they are reflex ovulators. The rabbit's pituitary hormones maintain a constant supply of ova in the preovulatory condition; then, upon coitus, a reflex originating in uterine and vaginal nerves triggers release of LH. Consequently, meiotic maturation takes place, followed by ovulation approximately ten hours after coitus. By this time, the sperm have been transported to the tubes and are ready for fertilization.

During the early 1930s, Pincus and his collaborators performed extensive studies of early parthenogenesis in the rabbit. They mated does to vasectomized bucks whose sterility had been carefully tested. In this way, they were able to trigger ovulation, while preventing fertilization. Some twelve hours later, they opened the animals' fallopian tubes and removed the chromosomally matured ova. These eggs were placed into culture (see chapter 8), and briefly exposed to different environmental insults. Over the years, Pincus found that parthenogenetic development could be triggered by many substances and conditions, including butyric acid, high salt concentrations, and brief exposure to either high or low temperatures. Of all techniques tested, cooling was the most efficient: more than 60 percent of exposed eggs initiated embryonic development.

This *in vitro* development usually did not proceed far, though. Most eggs developed only up to the two-cell, four-cell, or eight-cell stage, and then degenerated. However, a few eggs did get as far as the morula and blastocyst stages.

Pincus tried to carry development further. He transferred some of his early parthenogenetic embryos into the uteri of recipient female rabbits, and, in a few instances, liveborn baby rabbits were obtained.

Did these baby rabbits truly result from parthenogenesis? Most scientists would not accept such a conclusion, because there is no way to be certain that all the vasectomized male rabbits had indeed been fully sterilized. As a matter of fact, sperm were later discovered in the ejaculates of one of the males that had been used.

However, Pincus discovered an ace up his sleeve. By the late 1930s, preparations of human chorionic gonadotrophin had become available — the same HCG now in use to induce human ovulation (see chapter 8.) Therefore, Pincus no longer had to rely on a vasectomized male to release eggs from the ovary; a simple hormone injection would do just as well. Using HCG, he obtained matured eggs, treated them *in vitro* to initiate parthenogenetic development, and then reimplanted them into a rabbit's uterus. As a result, two live baby female rabbits were born to mothers who had never been exposed to sperm.

In 1940, Pincus and Dr. Herbert Shapiro further refined the technique. They induced ovulation with HCG; then, they exposed and cooled the fallopian tubes containing the eggs. Finally, they closed the rabbit's abdomen. From about two hundred eggs so treated, one female offspring resulted.

After this experiment, Pincus became interested in the sex hormones, and he shifted his efforts into the line of research which led to the development of the oral contraceptives. He never again tried to produce parthenogenetic mammals. Furthermore, during the thirty-five years since Pincus' last success, no other scientist has been able to obtain live-born young in a mammalian species by parthenogenesis. This lack of confirmation has led many biologists not to accept Pincus' claim, despite his absolutely unchallenged reputation for capability and integrity.

Since 1940, parthenogenetic development in mammals has been studied by many eminent scientists, including Drs. M. C. Chang of the Worcester (Mass.) Foundation, A. Tarkowski of Warsaw, W. K. Whitten of Bar Harbor, and C. R. Austin, A. W. H. Braden, and C. F. Graham of England. Most of their work has been done on the eggs of mice, rabbits, and rats, but hamsters, sheep, cats, ferrets, and other animals have also been studied. This research has both expanded the list of agents capable of initiating parthenogenetic development, and demonstrated that different agents are of different effectiveness in different animals.

During the 1960s, the advent of reliable techniques for counting chromosomes in the cells of embryos opened up new experimental possibilities in parthenogenesis. A team in Warsaw under the direction of Dr. Andrzej Tarkowski used an *in vitro* technique, whereby parthenogenesis was initiated in mice by briefly passing an electric current through the fallopian tubes after ovulation had occurred. During the ten days following electroshock, the scientists checked the animals' tubes and uteri. Approximately 50 percent of the eggs exposed to electric shock initiated parthenogenetic development. However, only 30 percent of the activated eggs developed far enough to implant, and by day eight, most of these had died as well. No embryo developed beyond day ten, the half-way point of mouse pregnancy. A particularly interesting finding, which had also been noticed in other experiments, was that most of the parthenogenetic embryos were both smaller than normal and developmentally retarded. These findings remain unexplained.

Tarkowski's studies also helped to define the chromosomal consequences of parthenogenetic activation. Remember that eggs are ovulated after having completed the first meiotic division, but before completion of the second. At that stage, normal eggs contain the replicated chromosomes of one member of each chromosome pair. If an egg is activated by sperm penetration (fertilization), meiosis proceeds to completion, with extrusion of half of each replicated chromosome into the second polar body. This leaves a haploid set of chromosomes in the egg to join with the haploid set of the sperm, to reconstitute in the embryo the normal diploid number for the species. Parthenogenetic activation, however, can produce four different chromosome situations (Fig. 9.1):

1. Those leading to haploid embryos
   a. formation of the second polar body, leaving the egg with one nucleus containing a haploid set of chromosomes. This is identical to the situation accompanying normal activation.
   b. "immediate cleavage." In this case, the egg does not produce a second polar body, but divides into two equal-sized cells, each containing a haploid set of chromosomes. Therefore, the result is a two-cell, haploid embryo.
2. Those leading to diploid embryos
   a. separation of the replicated chromosomes into two haploid nuclei without formation of the second polar body or any other division of the egg cytoplasm. The two nuclei may then recombine, much as do the nuclei of egg and sperm, under normal circumstances.
   b. chromosomal separation without cytoplasmic division as in 2a, but with immediate organization of the chromosomes into a single diploid nucleus.

It should be noted as well that haploid embryos may convert to diploidy through a single episode of chromosomal replication without concomitant cell division.

Spontaneous Parthenogenesis

By contrast to the experimental situation, considerably less knowledge exists regarding spontaneous parthenogenesis in animals. Unfertilized eggs of some marine invertebrates frequently develop into normal adults; in addition, drone bees normally arise from unfertilized eggs. Liveborn parthenogenones have been identified in poultry.

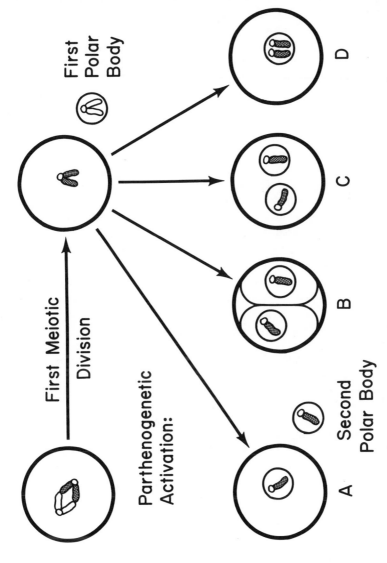

Fig. 9.1: The four chromosomal situations that can arise in an embryo after parthenogenetic activation.

However, in all these species, sex determination is different from that in mammals.

In mammals, with one known exception, spontaneous parthenogenetic development is unusual. The exceptional species is the golden hamster, where about 75 percent of all unfertilized eggs become parthenogenetically activated. However, development does not ever appear to go beyond the two-cell stage, and rarely does it progress even that far. In mouse eggs, the incidence of parthenogenesis varies with the strain of mouse, but is always under one percent. As in the hamster, any development is highly rudimentary.

Does natural parthenogenesis ever result in the birth of full-term offspring? This is difficult to answer, but to this date no verified case of spontaneous parthenogenetic birth has been reported in any mammalian species. Every now and then, a scientist will report the birth of young to an apparently isolated female laboratory animal, but it's usually difficult to definitely exclude the possibility of fertilization. Dr. W. K. Whitten of Bar Harbor noticed that low sex ratios (excessive numbers of females) were constantly recorded in certain matings in mice where the males had excessive numbers of defective sperm. Reasoning that parthenogenesis could provide one explanation for the abnormal sex ratios, Whitten performed genetic tests on the offspring and the parents. Since the female progeny were found to possess certain traits which could have been inherited only from the fathers, parthenogenesis was ruled out.

Parthenogenesis in Humans

To my knowledge, no work has been performed in experimental human parthenogenesis. Therefore, we may proceed directly to spontaneous human parthenogenesis.

Throughout history, indignant sniffs or snide snickers have greeted the claims of unmarried women that their babies must be parthenogenetic. However, in 1955, these claims received some sympathetic support from Dr. Helen Spurway, lecturer in biometry and eugenics at University College, London. In a talk entitled (unfortunately) "Virgin Births," Dr. Spurway reviewed what was then known of natural and experimental parthenogenesis, and offered the opinion that one human being in every 1 or 2 million might in fact be a parthenogenone.

How could such a claim be confirmed? Dr. Spurway explained that any parthenogenetic child would have to be female, that the parent-child blood groups would have to be compatible, and that the mother would be able to accept a skin graft from the daughter. (The reverse would not be true, since half of the mother's genes would have been cast off into the first polar body. Certainly, some of these "lost genes" would have been present in the mother in the heterozygous state. Consequently, although the daughter could have no gene not also possessed by the mother, the mother might have some genes not possessed by the daughter.)

Dr. Spurway's lecture was seized upon by the British tabloid *Sunday Pictorial.* Under banner headlines screaming "VIRGIN BIRTHS. Doctors Now Say: It doesn't always need a man to make a baby," the *Pictorial* issued a call for women to come forward and be tested if they believed they had had a child by "virgin birth." Nineteen women responded.

Of these nineteen, eleven were immediately disqualified because they had shown more erudition than the lexicographers by assuming that virgin birth meant conception through an intact hymen. Six more women were dismissed because their daughters were found to have blood cell proteins not present in the mothers. The eighteenth woman and her daughter had identical blood groups, but the mother had blue eyes, and the daughter, brown. Since the gene for blue eyes is recessive to the one for brown eyes, the mother could not have possessed a "hidden" gene for brown eyes. In the nineteenth mother-daughter pair, blood-group analyses and eye color were compatible with parthenogenesis. Three further types of genetic testing also failed to implicate a spermatozoan. Finally, reciprocal skin grafts were performed between mother and daughter. This test failed: both grafts were rejected. However, the physicians made the point that the daughter might in fact have had only maternal genes, but that subtle, unrecognized differences could have existed between mother and daughter in the manner of expression of the genes for tissue compatibility. Therefore, the authors concluded that they could not definitely disprove the mother's claim, and so, added another fascinating little conundrum to medical history.

Possible Applications of Parthenogenesis

Although at this time there appears to be relatively little in the way of practical use for parthenogenesis, the procedure is not totally frivolous. Biologists have long wished to have available a strain of haploid cells that could be grown in tissue culture. Since genetic heterozygosity could not exist in such cells, all mutations would be readily apparent, and so, the cells would be useful in studying spontaneous and induced gene mutation rates, as well as the effects of the mutations on the cells. Because some parthenogenetic embryos develop as haploids, it has been hoped that they might serve as sources for such cell lines. However, to date, there has been little success in culturing these haploid cells. One problem is that haploid cells are at a genetic disadvantage, since there is free expression of *all* the deleterious recessive genes which are normally masked in the diploid state by their dominant alleles. In addition, surviving haploid cells tend to become diploid as they grow, probably by replicating their chromosomes without undergoing the usual accompanying cytoplasmic division.

The great majority of diploid parthenogenones are homozygous for almost all their genes (Fig. 9.1). This is so because at formation of the first polar body during ovum maturation, one of each pair of replicated chromosomes passes into the polar body, thereby eliminating heterozygosity. Diploidization then occurs in the parthenogenones by division of the replicated single set of chromosomes *without* concomitant formation of the second polar body. The result is virtually universal genetic homozygosity in the parthenogenetic embryo. This homozygosity is usually cited as the reason why parthenogenones fail to develop far: the few deleterious recessive mutations that most organisms are thought to carry harmlessly in the heterozygous state now become homozygous, and cause the death of the embryo. This theory eventually may be tested by inducing parthenogenesis in the eggs of purebred mouse strains. These animals are thought to be homozygous for most of their genes, but with the lethal mutations having been bred out. Therefore, one might logically expect parthenogenetic embryos of these animals to develop normally to birth.

However, this probably will turn out not to be the case. Returning for a moment to Dr. Pincus, it will be recalled that he originally began to work with parthenogenesis in order to study certain aspects of the fertilization process. Basically, the question may be asked whether sperm penetration accomplishes anything more than contribution of chromosomes to the embryo. The answer is probably yes. For some time, it has been observed that the cytoplasm of parthenogenetic embryos appears abnormal. Then, in 1974, Dr. B. J. Gulyas of the National Institutes of Health published some electron micrographs of parthenogenetic rabbit eggs, which demonstrated abnormalities in cytoplasmic structures near the outer egg borders. The findings of this study were confirmed and extended by a similar 1974 report from Philadelphia's Wistar Institute of Anatomy and Biology. Hence, it may be concluded that entry of the fertilizing sperm triggers some cytoplasmic activities necessary to normal embryonic development.

We can see, then, that study of parthenogenetic embryos may yield valuable information about the fertilization process. This knowledge, in turn, may increase our understanding and improve our treatment of certain types of infertility. Conversely, such research might lead to the development of new contraceptives. In addition. scrutiny of later stages of parthenogenesis might tell us something about pre- and post-implantational early pregnancy wastage.

Teratomas are human tumors which consist of disorganized masses of embryonic tissue. They are believed to arise in a parthenogenetic fashion from egg cells. Some teratomas are benign, but others are highly malignant and kill their bearers. A better understanding of parthenogenesis in general might improve our ability to deal these deadly cancers.

At present, there does not seem to be any constructive way to directly apply parthenogenesis to human reproduction. Aside from any other consideration, the unmasking of deleterious recessive genes probably would preclude the use of parthenogenesis to effect (female) sex choice of offspring, for whatever reason.

Nicholas von Hoffman has expressed concern that a band of Amazons might decide to eradicate all the males in the world, and then reproduce parthenogenetically. Even if we disregard the problem of unmasked recessives, this scheme would not be likely to work. Most biologists believe that the recombination of genes incident to sexual reproduction is necessary for successful evolution. Even protozoa which usually reproduce by simple fission must intersperse their episodic asexual buddings with an occasional conjugation. Therefore, the Amazons, having slaughtered in haste, would find themselves at a genetic dead end, forced to rely upon mutations (the vast majority of which are harmful) as their only source of genetic variability.

10

Cloning

Cloning is probably one of the most futuristic and uncertain areas of Reproductive Engineering. This statement may be surprising, since the comments of many scientists and laymen might lead us to believe that mass human cloning will be with us before the turn of the century. We are told that then, sexual reproduction may become passé, or even forbidden. Asexual reproduction programs might carry the day, with directed breeding of just the right number and kinds of soldiers, policemen, dimwits, and geniuses necessary to carry on the work of Big Brother's society.

My own opinion is that all this talk represents no more than group hysteria, a mass construction of bogeymen. I do believe that cloning will one day be possible in humans, but I don't think that day will come soon. Nor do I think it likely that cloning will pose a significant threat to either individuals or society. On the contrary, the application of the concept may turn out to be of great value to us.

What Is Cloning?

In modern biological parlance, a *clone* consists of all the descendants arising from a single cell by asexual reproduction. This situation presents a major contrast to sexual reproduction, where the offspring are genetically different from both parents, due to recombination via haploid gametes of half the maternal and half the paternal genes. In asexual reproduction, however, meiosis and genetic recombination do not occur. Reproduction is accompanied by transmission to the progeny of the entire genetic complement of a single parent. Therefore, parent and offspring are genetically identical. In fact, except for any mutations which may occur, all members of a clone share a common genetic constitution.

Cloning is not an artificial tinker toy contrived by and for white-coated eccentrics; examples of it may be found in nature. Amoebas reproduce asexually, by periodic replication of chromatin, followed by division of the nucleus and cytoplasm. This mitotic process is identical to the division of a one-cell embryo into two cells, but the two newly formed identical cells separate and go their individual ways, rather than adhering to one another, eventually to develop into a single multicellular organism. Imagine, then, one amoeba in a tank of water. By the time this protozoan and its descendants had all divided ten times, there would be 1024 amoebas, all with a single ancestor, and all genetically identical. This is cloning.

Another example can be drawn from cell culture. When a sample of amniotic fluid is mixed with culture medium and placed in an incubator, the live fetal cells settle individually to the bottom of the plastic dish. Here they replicate, using the same mitotic process as the amoeba and the one-cell mammalian embryo. Within days, each original amniotic-fluid cell produces a colony of thousands or even millions of identical cells. Another word for such a colony is a clone, and the individual members of the clone, genetically identical and produced by asexual reproduction, are called *clonees.*

In contrast to amoebas and cultured cells, asexual reproduction is not part of the natural life cycles of mammals and other vertebrates. But the principles involved in cloning can be applied experimentally to these organisms. Frogs and mice have been produced by two laboratory techniques of asexual reproduction: nuclear transfer and embryo fission.

Cloning by Nuclear Transfer

Of central importance to this technique is the concept of cellular differentiation, or specialization. The cells of an early embryo are *totipotent:* that is, they have the capacity to give rise to any and all cell types. Such cells are *undifferentiated.* As an embryo grows and develops, different groups of cells specialize into various organs and tissues. A liver cell acquires the capacity to carry out the biochemical functions of the liver, and when it divides mitotically it forms two liver cells. It cannot transform itself either morphologically or functionally into a pancreas cell. Like most specialists, it has become good at doing some things and incapable of doing others.

The totipotent cells of the early embryo contain a complete set of genes in duplicate. These genes possess all the information necessary to the functioning of both the embryo and the mature organism. For years, an argument raged among biologists: does cellular differentiation involve the physical loss of genes, or does it simply entail selective activation and inactivation of different sets of genes? In other words, does the liver cell retain the full complement of genes characteristic for the organism, but only permit the "liver-function genes" to operate? Another way to put the question is to ask whether cellular differentiation is irreversible (as it would be if genes were lost or destroyed), or whether differentiation can be reversed, say by reactivation of genes inactivated during differentiation.

It has long been realized that cellular differentiation in plants and invertebrates is not quite as final as it is in mammals. Both halves of a transected worm will regenerate their missing parts. When a crab loses a claw in a fight, it simply grows another one. Similarly, the ability to grow entire plants from small cuttings is well known. In addition, true cloning has for many years been an accomplished fact in plants. Single cells can be removed from the root-tip area of carrots, and if maintained in proper culture media, many of them will develop into mature, complete carrot plants. Thus, the root-tip cell, nucleus and all, is capable of dedifferentiating into a state comparable to the totipotential seed, and then redeveloping into a complete plant. Since many root-tip cells can be removed from a carrot, a single vegetable can give rise to innumerable progeny.

Experiments such as these demonstrate that in some organisms, fully differentiated cells retain the capacity to undergo dedifferentiation and subsequent redifferentiation into various specialized tissues. This is fine for carrots and crabs, but mammalian cells don't seem to work that way. Dogs, cats, and people cannot regenerate missing body members, and no amount of manipulation has succeeded in persuading a cultured mammalian cell to do anything other than generate more of its fully differentiated kind — if it grows at all.

In 1952, however, Drs. Robert Briggs and Thomas King of Philadelphia had an idea. Proceeding under the assumptions that cellular differentiation does indeed occur by selective gene activation and inactivation, and that the crucial signals to the genes originate in the cytoplasm, they reasoned that by changing the signals, they might be able to change gene activation patterns, thereby altering cellular differentiation as well. To accomplish this feat, the obvious technique seemed to be to change the cytoplasm — or, better, to transfer the nucleus to the interior of a different type of cell.

Briggs and King began by removing nuclei from frog eggs. Working under a special dissecting microscope, they inserted fine glass needles into the eggs, drew out the nuclei, and discarded them. Next, the scientists separated frog blastula embryos (comparable to mammalian blastocysts) into their component cells; the same glass needles were then used to break open the individual cells and inject the freed nuclei into the previously enucleated eggs. More than 50 percent of these injected eggs initiated embryonic development, and many of them developed into tadpoles. About half the tadpoles appeared entirely normal, but the other half demonstrated a variety of morphologic abnormalities.

Thus, Briggs and King took the first step in cloning vertebrates. They demonstrated that a transplanted diploid nucleus, at least in some cases, could direct an egg to undergo normal embryonic development. True, blastula cells and their nuclei are still totipotent (undifferentiated), but nonetheless, these experiments established the principle of asexual reproduction in vertebrates.

During the early 1960s, the English biologist J. B. Gurdon built upon the groundwork laid down by Briggs and King, and succeeded in cloning frogs from differentiated cells. His procedure began with the removal of the nucleus from a frog egg, either by the glass needle technique or by subjecting the egg to ultraviolet irradiation. Then, he dissected a cell free from the intestinal lining of a tadpole, disrupted the cell, and, with a glass needle, injected the intestinal-cell nucleus into the egg.

The majority of Gurdon's treated eggs failed to undergo any embryonic development at all. A smaller number did initiate development, but the embryos were grossly abnormal, and soon died. Only a few eggs — between 1 and 2 percent — developed into entirely normal tadpoles, which then underwent normal metamorphosis into adult frogs.

Carrying the procedure one step further, Gurdon performed what he called "serial nuclear transplantation." He obtained intestinal-cell nuclei from one of his asexually produced tadpoles, and by transferring a number of these nuclei into enucleated eggs, was able to produce a second generation of cloned frogs. All these second-generation frogs were descended from a single original intestinal-cell nucleus, and so, were genetically identical.

Gurdon noted that as the degree of differentiation of the donor cells increased, there was a corresponding decrease in the proportion of nuclei capable of directing normal development of the eggs. In addition, he could only manage to clone frogs from tadpole nuclei; nuclei taken from adult frogs did not seem to work. Nevertheless, the cells from the tadpole intestine were thoroughly differentiated, both anatomically and functionally; therefore, the fact that they could, at least occasionally, direct the development of an entire embryo demonstrated the truth of the hypothesis that each specialized body cell carries a full genetic complement, and that the inactive genes are capable of being reactivated by specific cytoplasmic signals.

For a time, no explanation could be found for the low rate of normal embryonic development reported by Gurdon after nuclear transfer. Especially tantalizing was his observation that embryos belonging to a common clone usually demonstrated similar abnormalities, but that there was no interclonal resemblance in this respect. A solution to the puzzle was proposed in 1970 by Dr. Marie DiBerardino and her associates at the Medical College of Pennsylvania. This team of scientists proceeded from the previously established knowledge that upon transfer into the egg, the nucleus increases in size, probably because of passage into the nucleus of the cytoplasmic proteins which both trigger cell division and induce widespread genetic dedifferentiation. By studying frog nuclei during the first eight hours after transfer, Dr. DiBerardino ascertained that immediately prior to cell division, the chromosomes in most of the nuclei possessed regions of tightly coiled DNA; such areas characteristically are composed of functionally inactive genes. As the chromosomes proceeded to divide, these regions did not behave normally, but appeared to resist division, so that chromosomal breakage occurred, with resultant structural damage and loss of groups of genes. Dr. DiBerardino claimed that the embryonic malformations were attributable to these chromosomal anomalies. More specifically, she explained that the basic difficulty arose because eggs and early embryos normally have much higher rates of cell division than differentiated cells. Therefore, in most cases of nuclear transfer, the cytoplasmic signal for cell division "goes off" before the egg has been able to completely dedifferentiate the genes in the nucleus. As a result, the chromosomes are forced to divide prematurely, resulting in breakage. This explanation would account for Dr. Gurdon's finding that the more differentiated a cell had become, the less likely it was to be able to direct successful embryonic development upon transfer. It would also explain the fact that members of the same clone demonstrated similar malformations. Depending upon the specific break points in the chromosomes of a precursor cell, all of its clonal descendants would inherit identical chromosomal errors, which, in turn, would produce identical anatomic defects.

The choice of the frog egg by Briggs and King and by Gurdon was a particularly fortunate one. Compared to the eggs of mammals, amphibian eggs are very large; moreover, they are highly resistant to damage at microsurgery. Another advantage is that frog eggs normally develop in pond water, rather than within the body of the mother. This natural ectogenetic behavior greatly simplified the procedures necessary to permit embryonic development in the laboratory. And so, despite the fact that it has been more than ten years since Gurdon's cloned frogs jumped into prominence, no one has yet succeeded in cloning a mammal by nuclear transfer. However, some interesting preliminary work has been performed, which eventually may lead to this accomplishment.

In 1970, Drs. Roger Ladda and Richard Estensen of Walter Reed Army Institute of Research used a drug named cytochalasin B to remove nuclei from mouse body cells in tissue culture. Upon exposure of a cell to cytochalasin B, the nucleus moved to the periphery, producing a bulge at the cell surface. In some cases, the nuclei completely passed out of the cells. The enucleated cells were then placed in solution together with chick red blood cells, and inactivated Sendai virus particles were added.

Sendai virus is a relatively common pathogen, known to produce respiratory infections. However, when chemically inactivated, the virus retains the capacity to cause cultured cells first to clump together, and then to actually fuse. By this means, Ladda and Estensen were able to unite the enucleated mouse cells with the nucleated chick red blood cells, forming interspecific hybrids. Moreover, the chick nuclei, previously small and shrunken as is characteristic of inactive nuclei, swelled noticeably upon entering the mouse cells and then initiated the metabolic activities which characteristically precede DNA replication. However, neither the chromosomes nor the cells actually divided.

This line of investigation was carried a step further by Drs. K. K. Sethi and H. Brandis, of West Germany. In a sense, these researchers reversed the Ladda-Estensen experiment. They used cytochalasin B to induce extrusion of nuclei from cultured mouse cells, but then they centrifuged the tubes of culture medium so that the nuclei settled to the bottom. They recovered the nuclei, mixed them with inactivated Sendai virus, and added HeLa cells, a widely used strain of cultured cells originally grown from a biopsy specimen of a highly malignant human cervical cancer. As a result, they obtained a peculiar hybrid cell with two functioning nuclei: the original HeLa cell nucleus, and the transferred mouse cell nucleus. The fact that the mouse nucleus was indeed functioning within the HeLa cell was shown by the finding of characteristic mouse proteins on the surfaces of the new hybrid cells.

Although Dr. Christopher Graham of England has used similar cell-fusion techniques to produce hybrids of haploid mouse eggs and diploid mouse body cells, no one has yet succeeded in transferring a body-cell nucleus into an enucleated mammalian egg. Successful cloning of a mammal by nuclear transfer will have to await technical refinements. So far, mammalian cloning can be accomplished only by embryo fission.

Cloning by Embryo Fission

This procedure has been mentioned in several earlier chapters. A number of scientists, most notably Dr. Beatrice Mintz of Philadelphia, have succeeded in disrupting mouse embryos at the two-cell stage. This can be accomplished mechanically, by viewing the embryos under a dissecting microscope, and gently pulling the cells apart with tiny surgical instruments. Alternatively, the embryonic cells can be separated by exposing the intact embryos to certain chemicals which dissolve the biological "cement" that holds cells together. When the cells are separated (by either technique), and placed into a recipient mouse uterus, they give rise to two complete, genetically identical embryos, both of which stand a fair chance of being born alive. The procedure is not really novel, since it is the exact experimental equivalent of natural identical twinning. Large clones can be constructed by serial embryo fission, a process analogous to Gurdon's serial nuclear transplantation.

Four-cell mouse embryos also can be dissociated, and each cell can be transferred to a recipient uterus. The result will be artificial identical quadruplets. (Even this is not a true departure from nature: the fertilized egg of the armadillo almost always dissociates spontaneously at the four-cell stage, so that mother armadillo routinely counts on the appearance of quads.) Cells obtained from eight-cell mouse embryos frequently can be induced to develop into whole new mice, but when cells are taken from embryos beyond this stage, the procedure usually is not successful. Cells from morulas and blastocysts seem to have completely lost the ability to start all over from scratch.

It should be noted that there is a major difference between cloning by nuclear transfer and by embryo fission. With the former procedure, the cloner can apply a genetic value judgment in choosing his prototype. This is not possible, however, with embryo fission, unless the embryo happens to be from a highly inbred animal strain, or unless it was itself created by nuclear transfer.

Potential Applications of Cloning

From the standpoint of scientific achievement, the current state of cloning is considerably more primitive than that of preimplantational ectogenesis. Consequently, the apparent applications are less numerous and not as well defined. Nevertheless, as was done with ectogenesis, the applications may be broadly divided into those potentially useful in different areas of research, and those which might be directly applied to the clinical practice of medicine, especially with regard to reproduction.

Possible Research Applications of Cloning

The most obvious use of cloning is in the study of nuclear differentiation and dedifferentiation. It should be noted that this field may yield valuable information right from the present time. In fact, as previously stated, we have experimental cloning to thank for the knowledge that cellular differentiation is accompanied not by loss of genes, but by selective gene activation and inactivation; that this differential activation process is reversible; and that the determinants of gene activity reside in the cytoplasm. Many other basic and important questions concerning nuclear differentiation might be investigated through cloning. For example, precisely what is the cytoplasmic messenger-signal which enters the transferred, differentiated nucleus, and tells it to set its genetic material to work? How does this signal operate? What is the site on the chromatin at which the signal delivers its message? Answers to these general questions have the potential to crack wide open the puzzle of regulation of gene activity; this in turn would have tremendous implication regarding the understanding of normal embryonic differentiation and the birth defects which result from abnormalities in these processes. Furthermore, as explained in a previous chapter, any insight into the mechanisms of reactivation of dormant genes could well provide clues to both the origin and the treatment of cancer.

Related to research in nuclear differentiation is the study of cell division. Although the normal impetus to cell division probably is to be found in the cytoplasm, we don't know what it is, or how and at which site it exerts its influence. We don't know how some tissues (such as skin and the blood-forming elements) manage to maintain a rapid rate of cell division, while other tissues (e.g. brain), almost never divide. Experiments utilizing nuclear transfer might furnish information applicable to normal growth and development, healing of injuries, and, of course, cancer.

Even some of the mysteries of chromosomal anomalies might yield to the cloning approach. Work such as Dr. DiBerardino's might link embryonic chromosomal abnormalities, numerical or structural, to inappropriately timed cell division, and then elucidate the reasons for the poor timing. Such information might be applicable to the prevention of mongolism and like diseases, by suggesting a preventive method that could be preferable in many ways to amniocentesis and abortion.

Immunology is another field where nuclear transfer experiments might advance knowledge. Immunology deals with the antigen-antibody reactions that help fight off invading bacteria, cause hay fever, and bring about the rejection of transplanted body tissues. A cell taken from an adult animal, deprived of its nucleus, and given a nucleus from another animal might do some very interesting things. For example, which new antigenic genes would the cytoplasmic signal turn on? Would the nucleus of such a hybrid cell invariably produce antigens considered foreign by the donor of the cytoplasm? Following upon this line of thought, would the donor reject and destroy the hybrid cells if they should be reintroduced into his body? Research of this sort might permit us to learn how to both inhibit and enhance our immunologic activities to advantage. For example, it might be nice to be able to "turn up" our antibody systems during flu season, and "turn them down" when the pollens begin to float through the air.

Twin studies have long been profitable in medical research. Since identical twins have identical genetic endowments, they constitute the ideal setup for anyone wishing to study the relative determinative importance of heredity and environment in different physical and mental traits. Animal clonees, whether produced by nuclear transfer or embryo fission, could serve as subjects in investigations concerning conditions ranging from the genesis of birth defects to the development of personality traits; from drug sensitivity to intelligence.

Cloning as Directly Applied to Human Reproduction

With regard to the gathering of headlines, only ectogenesis has outdone the subject of human cloning. Numerous bizarre scenarios have been put forth, based on the misuse of cloning by tyrants and madmen. However, there is ample reason to suspect that the proposed hazards of "the clonal man" have been greatly exaggerated. The same general counterargument that was applied to the theoretical dangers of ectogenesis may be used in the cloning situation: slavery must first be established before cloning can be applied as part of a directed reproductive policy. In addition, several specific arguments can be made.

Let's examine some of the imagined future applications of cloning. It has been suggested that a dictator might want to clone himself extensively, thereby spreading his "carbon copies" throughout the domain. This, however, seems unlikely: the last thing that any tyrant wants is competition. On the other hand, suppose a despot were to decide to clone a single child to carry on after his own death? This reproductive technique would seem to offer him no particular advantage. Would the Macedonians have been worse off under a cloned Philip than under Alexander?

Cloning might, in theory, be used to create large numbers of people to different specifications. In this way, clones of dull, unintelligent persons might be created to do tedious, unpleasant tasks. Large armies of soldiers could be cloned from the nastiest thirty-year master sergeants in the country. It has even been suggested that clones might be established from genetically legless persons, to create astronauts who would fit better into space capsules. However, these schemes all presuppose a nationally directed reproductive program, where having children is no longer an individual prerogative, and where a person's destiny is foreordained. Insistence upon freedom of individuality probably is the best safeguard in this area. Such insistence will be far more valuable than any laws that might be passed to forbid the practice of cloning — bans that no self-respecting dictator would feel compelled to obey, anyway.

A problem for our hypothetical dictator would be that cloning is an inefficient technique. Just like any other babies, clonees have to grow up, and an army of cloned master sergeants in diapers would not be helpful in doing battle with the Chinese next week. To obtain his armies, his morons, or his astronauts, a dictator would be far more likely to resort to more immediate techniques. such as brainwashing, drug treatments, surgery, or even good old-fashioned orders at the point of a gun.

H. J. Muller foresaw great possible use for cloning in his positive eugenics ventures. Whether by governmental edict or by individual parental choice, this would involve the cloning of geniuses in accordance with need in different areas of human endeavor. We could create teams of Einsteins, Newtons, Mozarts, or Chaucers.

Even if we pay no attention to the arguable question of whether genius by committee would in fact constitute a blessing, there still exist strong objections to such a plan. For one thing, it ignores the considerable contribution to a person's functioning made by environment. Although members of a common clone would be genetically identical (except for mutations), it is unlikely that the overall resemblance between themselves and their genetic parent would be any greater than that which normally exists between identical twins. Twins do bear strong resemblances to each other, but they are not "carbon copies," even when they are brought up together, and therefore presumably share an identical postnatal environment. Even prenatal environment is important: it is usually assumed that twins are exposed to identical intrauterine conditions, but this is not so. For example, in identical twins — with identical genes — one twin may be born normal and the other anencephalic. This may be explained by differential placental blood supplies, which, in turn, would cause one fetus to receive less oxygen, or more environmental teratogen, than the other. Clearly, no men are truly created equal. The extreme difficulty of producing exact copies of human beings is explained with grace and wit by Dr. Lewis Thomas in his article "On Cloning a Human Being," which is included in the reference list at the end of the book.

Another problem with cloning for positive eugenic purposes is based on the observation that greatness so often seems to be simply a matter of the right man happening to be in the right place at the right time. It could even be argued that environmental and social conditions continually select for genius, and so, by limiting our genetic diversity through cloning (as would also occur with parthenogenesis), we might find ourselves with fewer, not greater, numbers of geniuses, being saddled instead with groups of Xeroxed has-beens. Limitation of genetic diversity generally seems unwise.

Another serious potential problem related to cloning concerns the psyche of the clonee. Whether he were cloned from his de jure parent or from a particularly illustrious person, a cloned individual would be very likely to have his performance measured in the light of unreasonable expectations. In fact, I would wonder whether any sort of comfortable parent-child relationship could be formed under these circumstances. In this context, it might be appropriate to recall an aspect of the early life of Vincent van Gogh. His parents had had an older son, whom they also had named Vincent. This boy had died before the birth of the future artist. The second Vincent later described the discomfort and anxiety he had felt each day while walking to school, because his path led through the graveyard, where he could see his brother's tombstone with *his own* name on it. Whether this experience contributed to van Gogh's famous psychosis is problematic, but the possibility should give pause to would-be cloners of humans.

It has been suggested that cloning would present us with the ability to keep genetically identical copies of ourselves deep-frozen, so that we would have available a ready supply of "spare parts" that could be transplanted to ourselves whenever necessary. Certainly, this situation would present a raft of novel legal and ethical dilemmas, which, as usual, can be scarcely imagined, let alone solved, at this time. I suspect that most persons at present would find such an arrangement highly repugnant, at least until such time as they were to find themselves in mortal need of a compatible organ for transplantation.

A less grisly means of supplying anatomic replacements would involve developing the ability to grow specific body parts from single cells, for example, from cells obtainable by skin biopsy. Right now, such an eventuality must be classified as science fiction, but it certainly could one day become a reality. Before it does, though, giant strides would have to be made in tissue culture techniques. In addition, we'd have to learn the intricacies of turning on the genes for development and function of a desired organ, while turning off all other genes, including those that were functional in the donor tissue.

Theoretically, cloning might be used for sex choice of offspring, either in carriers of X-linked diseases, or for family planning. The potential problems associated with having your identical twin for a parent, however, would seem to make this technique less appealing than the procedures described in chapters 7 and 9: preconceptual sperm separation, and *in vitro* fertilization with preimplantational sexing of blastocysts.

Right now there exist tremendous practical obstacles to any proposed application of cloning. Cloning is conceptually simple, but technically difficult. Considerable experimentation will be necessary before the procedures will be able to be performed on the tiny, fastidious eggs of mammals. Furthermore, it is necessary to further elucidate the problem of malformations in the clonees due to chromosomal errors. Means of preventing these defects will have to be established.

In addition, at this time and in the foreseeable future, there appears to be no rational excuse whatever to clone a human being, and I really do not expect that biologists will try to do it. If the technical problems can be solved, it is much more likely that cloning will be applied to increase our food supplies, through rapid mass-production of high-quality grains, vegetables, and food animals.

However, can we be certain that there will forever continue to be no reason to clone humans? It would seem foolhardy and perhaps even presumptuous to claim that our descendants, living in a different world from our own, will certainly have no reason to clone humans. They may well develop uses for the procedure that cannot at this time be imagined. And if they do, it is not our concern.

It would be inappropriate at this time to react with hysteria and try to forbid all research in the field of cloning. These basic experiments can yield information equally applicable to the production of slaves or the prevention of birth defects and cancer. We cannot with assurance ban the one without denying ourselves the other.

11

The Synthesis of Life
From Nonliving Matter

We have considered various means of producing organisms by generative procedures other than the usual for the species. All these techniques have involved the manipulation of already living material. In the present chapter, however, I will give some thought to the possibility of the true creation of life, the synthesis of a living organism from nonliving components.

How does one define a "living organism"? There really is no fully satisfactory answer, but a reasonable approximation would be to say that living organisms possess the abilities to grow, to reproduce, and to provide for their own energy needs by the use of metabolic reactions.

The road to the *de novo* synthesis of living matter is going to be a long and tortuous one. Molecular biologists are only now beginning to develop techniques that may be of use in the field. With this in mind, let us consider the small number of pertinent scientific accomplishments to date and then think about their theoretical ramifications and possibilities. The discussion proceeds from the construction of simpler to that of more complex organisms. I'll start with the synthesis of DNA (genes and viruses), then go on to prokaryotic cells, and finally consider eukaryotic cells, both protozoal and metazoal.

Synthesis of DNA (Genes and Viruses)

This is both a possible end unto itself, and a necessary first step in the attempted synthesis of complex organisms. As has previously been mentioned, individual genes consist essentially of short lengths of DNA, and viruses are nothing more than a few genes joined together and surrounded by a protective protein coat. Most scientists consider viruses to inhabit a "gray zone" between the living and the nonliving worlds. They do not metabolize, nor do they grow. They may be said to have achieved the Playboy ideal, in that they are composed solely of reproductive material, and their only apparent function is to reproduce.

Individual genes have already been synthesized. In 1970, a team headed by Dr. Har Gobind Khorana of M.I.T. succeeded in assembling two genes, starting from scratch. First, the scientists put together a yeast gene which codes for the production of a substance necessary for the incorporation of the amino acid alanine into proteins. Khorana and the other scientists accomplished this feat in three steps. First, using ordinary shelf chemicals and techniques of organic chemistry, they synthesized fifteen relatively short lengths of DNA. Next, with the help of an enzyme that joins DNA together, the team assembled the fifteen small subunits into three larger ones. Lastly, another enzymatic procedure was employed to join together the three large subunits, thereby forming the single gene. A couple of years later, Khorana's team used similar techniques to assemble a longer gene from the bacterium *E. coli*. This gene directed the production of a cellular material without which the amino acid tyrosine could not be incorporated into proteins.

An attempt to copy a gene can be made only after having first ascertained the specific basic structure of the DNA composing the natural gene. How this is done will not be discussed here, but references on the subject will be included at the end of the book.

Unfortunately, Khorana's synthetic genes could not be put to a functional test. The basic composition of the associated regions of DNA which control gene function was unknown; therefore, the synthetic genes could be neither turned on nor shut off. A first step toward gene control was reported in 1975 by a group headed by Dr. Robert C. Dickson. Also working with *E. coli*, this team succeeded in analyzing the structure of the region of DNA which regulates the function of the genes for the production of enzymes necessary in the metabolism of galactose.

This type of research opens the way to construction of a virus to specifications. During the past decade, both Dr. Sol Spiegelman and Dr. Arthur Kornberg managed to synthesize small viruses. However, both accomplishments were more a matter of replication of existing DNA than totally new synthesis of the material. Furthermore, in neither case was the virus "made to order" by assembly of specifically desired genes.

Why would anyone want to create a new virus? The most obvious answer was discussed in chapter 4. It would be nice to be able to make up a virus composed solely of the gene to permit integration into the host chromosome and the gene which would correct a genetic disease, such as hemophilia. This might give us a one-shot cure for different genetic disorders.

"Therapeutic viruses" might not be of use only to those suffering from genetic diseases. They might also be used to introduce new genetic traits into "normal" human beings. For example, we might wish to acquire genes that would permit us to digest, absorb, and utilize cellulose, which now passes unchanged through our intestines. Or, we might acquire genes that would confer upon us resistance to disease-causing microorganisms.

If this type of genetic therapy were to be applied to individuals after their birth, it would benefit only the persons treated. If, however, the virus were inoculated into preimplantational embryos *in vitro*, presumably all the embryonic cells could be altered, including those that will later give rise to the germ cells. Hence, all descendants of any treated embryo would inherit the new genetic characteristic.

Before such a plan could be instituted, however, it would have to be tried extensively in laboratory animals, to be certain that no harmful side effects would accompany such gene transplants. The first potential side effect that comes to mind is, of course, the induction of cancer by viruses.

Synthesis of Prokaryotic Cells

It is difficult enough to construct DNA to order, but it will be even more troublesome to construct an entire cell. Basically, there are two types of cells: prokaryotic and eukaryotic. Eukaryotes are the more complex cells: their hereditary material is organized into chromosomes which exist in a membrane-bound nucleus; and the various machinery for cellular metabolism is arranged in well-organized membrane-surrounded compartments in the cytoplasm. In prokaryotes, on the other hand, the DNA is a simple circular molecule, and the metabolic machinery is loosely distributed throughout the cytoplasm. Prokaryotic cells include bacteria and the simpler algae; the rest of the organisms on earth are eukaryotes.

How can one begin to go about synthesizing a living cell, even a relatively simple prokaryote, with all its mysterious cytoplasmic components? Dr. Howard J. Morowitz, a molecular biophysicist, suggests we might proceed according to the same general plan used in synthesizing genes: determine the structure, and then, using techniques of organic chemistry and enzymology, copy it. But how to determine the structure of a living cell? Morowitz has an idea. It has to do with temperature.

Most people know about the Fahrenheit and Centigrade scales for measuring temperature. But there is also a lesser-known third scale: the Absolute. Absolute zero is defined as that temperature at which all molecular motion ceases; it is 273 degrees below zero Centigrade.

Absolute zero is so far a theoretical concept, but temperatures within a couple of degrees of it have been reached. Furthermore, when microscopic plants and animals are maintained for a while at these extremely low temperatures and then are thawed, they resume function, with no evidence of having been damaged.

Therefore, Dr. Morowitz suggests that we might be able to maintain cells near Absolute zero long enough to use complex and sophisticated molecular biological methodology to discern their atomic and molecular structures. Then, with chemical and enzymatic techniques, the structures might be copied. Finally, in the same manner that microscopic creatures resume molecular motion upon thawing, Dr. Morowitz believes that the synthetic cells would initiate activity, in effect, "setting themselves into motion."

All this is theoretical, but it is a fact that we already have the technical knowledge necessary to construct any cell component. The problem is that we don't know the basic structures and relationships of the different parts of cells.

Dr. James Danielli of the Center for Theoretical Biology in Buffalo has suggested several reasons why we might want to construct prokaryotes to order. Among these are the potential advantages of being able to design new organisms that could help us in dealing with our environment. For example, we might come up with a bacterium that could desalinate sea water, or provide extremely efficient sewage treatment. Perhaps the organisms might even produce useful metabolic by-products. Another possibility would be to produce organisms capable of manufacturing large quantities of necessary biologicals, such as antibiotics, antibodies, and hormones.

Synthesis of Eukaryotes: Protozoa

The simplest eukaryotic organisms are the protozoa, animals composed of a single cell. In contrast, metazoa are animals made up of more than one cell. Probably, the best-known protozoa are amoebas, but there are innumerable other varieties of one-celled creatures to be found in our oceans, rivers, lakes, and ponds.

Amoebas can be reconstructed. Cellular fragments have been taken from several different amoebas and injected together into an empty amoeba cell membrane, or "ghost." This manipulation will produce a normally functioning single amoeba. Although this obviously does not constitute new synthesis of a living organism, it does illustrate that, at least in amoebas, precise placement of cell components is not necessary to subsequent cell functioning.

What has been said regarding the technical aspects and potential uses of prokaryotes would seem to apply equally well to the construction of protozoal eukaryotes. An additional factor that should be borne in mind is the fact that eukaryotes are more complex cells. Specifically, little is known about the protein-and-lipid membranes that surround the chromosomes and the cytoplasmic metabolic machines. In addition, the chromosome itself is a far more complicated structure than the simple circular molecule of DNA seen in prokaryotes.

Synthesis of Eukaryotes: Metazoa

From the sensational point of view, what we have so far discussed is small potatoes. The story of the creation of bacteria and amoebas would describe a tremendous scientific feat, but it wouldn't sell many newspapers or magazines. A synthetic metazoan, however, would be a different story.

We'll divide this part of the discussion into the synthesis of nonhuman metazoa, the synthesis of humans, and the synthesis of superhumans. Plants will be included with the first subcategory, even though, not being animals, they are not metazoa. However, since they are multicellular and eukaryotic, it is convenient to discuss them at that time.

At present, there are no scientific accomplishments specifically pertinent to the synthesis of metazoa, and so, the discussion will be entirely theoretical.

Synthesis of Nonhuman Metazoa and Plants

Fictional treatments of the synthesis of living organisms, from *Frankenstein* on down, have involved the creation of the full-grown organism in an inanimate condition, after which the spark of life is somehow implanted in it. Undoubtedly, such plans represent a sort of literary fundamentalism. However, a biologist would consider them to be extremely inefficient methods of producing metazoans. To construct a metazoan from scratch, it would be far more reasonable to proceed from the fact that many-celled organisms originate in one cell. Hence, a scientist probably would construct a pair of germ cells, or a single cell which could divide repeatedly, thereby permitting the metazoan to literally assemble itself, after the first step.

In the construction of metazoa, there arises a totally new problem. With increasing numbers, there is also increasing cellular specialization, or differentiation. Therefore, the cells not only must divide, but must also differentiate properly. And, as has been mentioned, even the most basic aspects of differentiation remain nature's secrets.

Therefore, at first, an aspiring creator of metazoa probably would have to be content with making copies. Using Dr. Morowitz' Absolute-zero technique, he might be able to discern the structure of a cow egg and a bull sperm, copy them, fertilize the egg with the sperm, and then either implant the embryo into a cow uterus, or, if then possible, develop it ectogenetically.

However, the basic question arises: why in the world would anyone want to go to all that trouble just to reproduce a cow? Ordinary bovines already are quite freely available in nature. The only possible reason I can conceive of for pursuing such a line of research would be to establish the technical aspects of the science. Once copying were to become possible, with further related research, we might be able to move on to creative creation: the manufacture of new species.

For example, if we had the means to control an organism's differentiation, as well as to construct its original cell, we might create a food animal with incredibly nutritious meat, an animal which used the waste material of other organisms as its own food. Or, we might design a food plant that was capable of growing in the desert — or a plant capable of taking up and utilizing free nitrogen and phosphorus from the soil, thereby eliminating the need for fertilizer. Would you like to add a few items to the list? It's a game anyone can play. Right now, the whole business is thoroughly visionary, but by the time we've sorted out the intricacies of cell structure, embryonic differentiation, and ectogenetic technology, we may well have figured out enough of the genetic code to be able to produce organisms to specification. I say "we," but of course I mean our distant descendants. I think that before any of this becomes possible, even the most famous of us will be no more than shadowy names in esoteric reference books.

Synthesis of Humans

Here, we come to consider the *Frankenstein* story, the great concern of many of the would-be prophets who plead for the passage of legal prohibitions against genetic research "before it's too late." Yet, as I see it, there is no potential problem here at all.

As described, at any time in the future, the scientific knowledge and the technological facilities necessary to create a human would be overwhelming. Therefore, this would not be a feat possible of accomplishment by the lone mad scientist, working in his cellar. And no group of legitimate scientists could possibly see this as a worthwhile goal.

Create a specimen of *Homo sapiens* from inorganic matter? Why bother? There is already more than a plentiful supply. You want a team of slaves or lackeys? It's easier to brainwash a group of naturally produced humans, if need be, from the cradle. In addition, it's highly probable that before we will be able to synthesize real humans, we will have the capacity to turn out robots, mechanical men who will be able to carry out distasteful or dangerous work, and who will generate no moral or ethical anxieties.

Synthesis of Superhumans

In that far-distant time when we have completely cracked the genetic code and have mastered all the techniques of genetic and reproductive engineering discussed in this book, we may then properly address ourselves to the question of producing supermen. Until that time, genetic efforts will be directed only toward the amelioration of disease.

All the heretofore-proposed schemes for breeding supermen pale before the thought of truly creating intelligent beings to exact specifications. Since we possess at least hundreds of thousands of gene pairs, the total number of ways in which they might combine is truly astronomical. Therefore, it is highly likely that the best gene combinations — whatever they may be — have never existed in any person ever born. Suppose we could not only construct a set of chromosomes with just the right combinations of the genes already present in humans, but also add a few extra genes to permit even better performance? What manner of marvelous mental and physical prodigy might we create?

Your guess is as good as mine. Suffice it to say, the most common reaction to such a thought is a negative one. One objection is that the act would be blasphemous. In other cases, the negativism represents nothing more than jealousy, the same sort of emotion involved when your neighbor happens to inherit a million dollars. But I think by far the greatest factor is fear. If we were to create such an awesome being, might not it and its kind turn on us, enslave us, and use us in *their* experiments?

Logically, I'd think not. Such admirable specimens of humanity would be expected to contain a large measure of decency, and to demonstrate proper love and respect for their creators. Our fear is based on projection of the manner in which we ourselves might behave in a similar circumstance.

And this being so, how would we be likely to construct safeguards if we were to decide to create a race of superhumans? Before we achieve that ability, I think we will have conquered space as well; therefore, we might decide to set our creations down on a separate, uninhabited planet, far from ourselves. First, of course, we'd have to stock the planet with appropriate plants and animals, but we should have no problem in creating these. Then, to make certain that our creations keep in line, we might furnish them with a strict code of arbitrary rules for living. Certainly, we'd not want them to taste the fruit of the tree of knowledge; that would be extremely dangerous for us. Furthermore, they might be told that those who had led exemplary lives would be permitted to spend eternity in the company of their benevolent creators, whom they had worshipped unquestioningly during their lifetimes. By such a plan, we might be able to observe at long range the progress of our extraordinary creations, at least until we had satisfied our curiosity, or until their maintenance became too much trouble. At that point, we could abandon them to their own devices, and move on to other experiments.

I have indulged in a little clearly labeled science-fictionalizing. In truth, there is nothing else that can be written about *de novo* creation of metazoa. Perhaps, one day far beyond our present intellectual horizon, after we have slowly and painfully acquired the necessary basic skills, this field will move from the realm of science fiction to that of science.

Notes

Preface

1. Genetic engineering: reprise. *Journal of the American Medical Association* 220 (1972):1356.
2. L. Kass, The new biology: what price relieving man's estate? *Science* 174 (1971): 786.
3. R. Sinsheimer, Genetic engineering: the modification of man. *Impact of Science on Society* 20 (1970):291.
4. J. Lederberg, Genetic engineering, or the amelioration of genetic defect. *Pharos* 34 (1971):10.
5. L. Thomas, Notes of a biology-watcher: on transcendental metaworry (TMW). *New England Journal of Medicine* 291 (1974): 779.
6. C. Stern, Genes and people. *Perspectives in Biology and Medicine* 10 (1967):522.

Chapter 1

1. F. C. Fraser, Genetic counseling. *American Journal of Human Genetics* 26 (1974):637.

Chapter 2

1. J. F. Crow, Discussion of paper by R. S. Morison. In *Ethical Issues in Human Genetics,* ed. B. Hilton, D. Callahan, M. Harris, P. Condliffe, and B. Berkley (New York: Plenum Press, 1973), p. 217.
2. Anonymous, The new eugenics. *Lancet* 1 (1971): 752.
3. W. E. D. Stokes, The right to be well born or horse breeding in its relation to eugenics. Quoted in W. E. D. Stokes on eugenics. *Eugenical News* 2 (1917): 13.
4. N. von Hoffman, *Scientists Are Awaiting the Birth of Little Invit with Horror.* Washington Post-King Features Syndicate, 1972.

Chapter 6

1. M. Twain (1882), On the decay of the art of lying. In *The Complete Humorous Sketches and Tales of Mark Twain*, ed.C.Neider (New York: Doubleday and Co., 1963), pp. 503-508.

2. S. J. Kleegman, Practical and ethical aspects of artificial insemination. In *Advances in Sex Research*, ed. H. G. Beigel (New York: Harper and Row, 1963), p. 114.

3. H. Brewer, Eutelegenesis. *Eugenics Review* 27 (1935):123.

4. Ibid., p. 125.

5. H. J. Muller, Significance of artificial insemination in relation to practical genetics in man. In *Advances in Sex Research*, ed. H. G. Beigel (New York: Harper and Row, 1963), p. 121.

6. J. M. Smith, Eugenics and utopia. *Daedalus* 94 (1965):494.

Chapter 8

1. M. S. Evans, The new totalitarians. *Private Practice* 6 (1974):20.

2. H. A. Smith, *Two-thirds of a Coconut Tree* (Boston: Little, Brown and Co., 1962), p. 110.

Glossary

ACTIVATION (of oocytes): The causing of oocytes to initiate embryonic development. Usually accomplished by the fertilizing spermatozoan, but can also frequently be done by any of a large number of artificial stimuli, in which case development is parthenogenetic.

ALLELES: The two members of a particular gene pair, located at corresponding points on homologous chromosomes.

ALPHA-FETOPROTEIN: A protein normally produced by the fetus and also by certain cancers. It is found in high concentration in the amniotic fluid when the fetus suffers from anencephaly or spina bifida.

AMNIOCENTESIS: Insertion of a needle into the amniotic cavity, for the purpose of withdrawing a sample of the amniotic fluid.

ANEUPLOID: A cell or an individual with an abnormal number of chromosomes.

BURDEN: A measure of impact of genetic disease, combining longevity and severity of manifestation. Burden may be thought of in terms of society, the family, or the individual patient.

CARRIER: An individual possessing a particular gene. Usually applied to the healthy, asymptomatic person who bears in single dose a (recessive) gene which causes disease only in double dose.

CHARACTER: The observable manifestation of the action of a particular gene or group of genes (syn.: *trait*).

CHIMERA: An individual composed of cells originally derived from two or more separate fertilized eggs.

CHROMOSOME: A linear aggregation of genes to form a chain of DNA and protein. Chromosomes are located in the cell nucleus. The number of chromosomes is characteristic for a given species: 46 (23 pairs) in man. Two of the 46 are sex chromosomes (XX in females, XY in males), and the other 44 are autosomes, carrying genes for general body functions. The members of a given pair of chromosomes are called homologous chromosomes.

CLONING: Asexual reproduction in which all the descendants (clonees) making up a clone are derived from a single cell, and are genetically identical.

CODE (verb): Direct the production of, as a gene for an enzyme.

CONGENITAL DISEASE: An illness or deformity which is present at birth.

CONSANGUINEOUS: Having one or more progenitors in common; therefore, genetically related.

CRYOBANKING: Cold-storage of biological material. Specifically used in this book to refer to the storage of frozen sperm or eggs.

DELETION (gene): An error in chromosomal replication during meiosis, which results in breakage and loss of a chromosomal fragment, thereby causing the chromosome to have a deficiency of genes.

DIPLOID: A cell or an individual with (the normal) two sets of chromosomes. Cells with three sets of chromosomes are triploid; four sets, tetraploid.

DOMINANT GENE: A gene whose trait will be expressed in single dosage, not being masked by its allele.

DUPLICATION (gene): An error in chromosomal replication during meiosis, which results in a chromosome's having a portion of its length copied in tandem, thereby coming to possess an excess of genes.

DYSGENIC: Tending to worsen the genetic endowment of the species (ant.: *eugenic*).

ECTOGENESIS: Embryonic development outside the body of the mother.

EMPIRIC RISK: A risk figure based on statistical population studies, rather than on knowledge of precise genetic inheritance mechanisms.

ENZYME: A biological catalyst which mediates a specific cellular metabolic reaction. The complete enzyme (holoenzyme) may be composed totally of protein, or may consist of a protein moiety (apoenzyme) which is attached to a coenzyme, often a vitamin or a mineral.

EPISOME: In bacteria, bundles of genetic material in the cytoplasm, separate and distinct from the chromosome.

EUGENICS: Genetic manipulations designed to improve the general health and quality of the human race.

EUKARYOTIC CELL: The more complex of the two cell types. There is a membrane-limited nucleus, containing definitive chromosomes, and the metabolic machinery is arranged in well-organized membrane-surrounded compartments.

EUPHENICS: Total biological and technological engineering, designed to help everyone realize his full potential for health and accomplishments. (Lederberg)

EUPLOID: Having the normal number of chromosomes for the species.

EUTELEGENESIS: "Reproduction from the germ cells of individuals between whom there is no bodily contact," as applied to "the eugenic breeding of man." (Brewer)

EUTHENICS: Environmental manipulations designed to improve the well-being of persons and groups.

FAMILIAL DISEASE: A disease which occurs in multiple family members. Often genetic in origin, but may also be infectious.

FETOSCOPE: An optical instrument, currently experimental, which is designed to be inserted into the amniotic cavity for the purpose of viewing the fetus and placenta. Also called *amnioscope*.

FITNESS (genetic): A measure of the ability of an individual to reproduce.

GENE: The basic unit of inheritance; a specific length of DNA which codes for the production by the body of a specific protein.

GENETIC DISEASE: An abnormal trait or traits, impairing health, usually caused by either a gene mutation or abnormal numbers of genes in an individual.

GENETIC ENGINEERING: The art or science of making practical application of the knowledge of the pure science of genetics, as employed in attempts to modify the structure, transmission, expression, or effects of genes.

HAPLOID: Containing a single set of chromosomes, as in the mature sperm or egg. In man, the haploid number of chromosomes is 23.

HETEROZYGOUS: Referring to the two members of a particular allelic gene pair: not identical; coding for different proteins (and therefore potentially capable of producing different traits).

HOMOZYGOUS: Referring to the two members of a particular allelic gene pair: identical; coding for the production of the same protein (and therefore producing the same trait).

HORMONE: A compound produced by a particular tissue, which is transported by the blood stream to other tissues, where it produces a characteristic biologic effect.

IMMUNE SYSTEM: Cells located in various body organs, which are specialized to produce antibodies for the purpose of destroying material (usually protein) recognized as foreign to the body.

IN VITRO: Literally, "in glass." Refers to biological processes which occur under laboratory, rather than natural conditions.

INBORN ERRORS OF METABOLISM: A group of genetic diseases, inherited in recessive fashion, and characterized by specific enzyme deficiencies which block specific steps of cellular metabolism.

INDEX CASE: The family member in whom a genetic disease is first recognized (syn.: *propositus*).

KARYOTYPE: A photograph of a set of chromosomes from a cell, arranged for study in a standardized fashion.

LINKAGE: Location of two or more genes close together on the same chromosome.

LIVING: Having the capacity to grow, reproduce, and provide necessary energy by the use of metabolic reactions.

LYSOGENY: The state of apparent peaceful intracellular coexistence between a temperate virus and its host.

MATURATION (of the oocyte): The meiotic process which takes place in the egg just prior to ovulation, and which is necessary for fertilization.

MEIOSIS: The process of reduction division by which mature spermatozoa or eggs come to contain half the number of chromosomes characteristic for the body cells of the species.

METAZOA: Animals composed of more than one cell.

MITOSIS: The process of replication and division of chromosomes; an integral part of cell division.

MONOGENIC: Traits determined by one gene or a single allelic gene pair.

MONOSOMY: The state in which a cell or an individual has one, rather than two, of any specific chromosome.

MULTIFACTORIAL: Traits determined by interaction of multiple gene pairs with environmental factors (often used loosely as a synonym for *polygenic*).

MUTATION: A structural alteration in the DNA of a gene, which causes the gene to code for the production of an altered protein.

NONDISJUNCTION: Failure of a homologous pair of chromosomes to separate during meiosis, causing the egg or sperm to either contain an extra chromosome or be a chromosome short.

PARTHENOGENESIS: Development (usually rudimentary) of an unfertilized egg into an embryo, fetus, or individual. *(See* ACTIVATION.)

PEDIGREE: A genealogical relationship diagram, used by geneticists in research and counseling. A "family tree."

PHARMACOGENETICS: The branch of genetics dealing with the variable manner in which individuals react to drugs and chemicals.

PLEIOTROPY: The ability of a single gene to have more than one effect upon an organism.

POLYGENIC: Traits determined by the action of more than one gene pair. *(See* MULTIFACTORIAL.)

PROKARYOTIC CELL: The less complex of the two cell types. The

DNA is a simple circular molecule, and the metabolic machinery is randomly distributed in the cytoplasm.

PROPOSITUS: The family member in whom a genetic disease is first recognized (syn.: *index case*).

PROTOZOA: Animals composed of a single cell.

RECESSIVE GENE: A gene whose trait will be expressed only in double dosage; when present in single dosage, its effect can be masked by a dominant allele.

REPRODUCTIVE COMPENSATION: Replacement by parents of genetically unfit offspring with fit ones.

REPRODUCTIVE ENGINEERING: Modification of the natural reproductive process; in humans, any generative maneuver other than sexual intercourse, usually involving manipulation of sperm, eggs, or embryos.

SENSITIZATION (immunologic): The first exposure of an individual's immune system to a particular foreign protein (antigen) which results in the production of antibodies to that protein.

SEX CHROMATIN: The characteristic appearance of the second X or the Y chromosome in the nuclei of specially stained body cells. A rapid screening technique for chromosomal sex.

SEX RATIO: In a given population, the number of males divided by the number of females, multiplied by 100.

TRAIT: The observable manifestation of the action of a particular gene or group of genes (syn.: *character*).

TRANSDUCTION: The introduction by a virus of new DNA into a bacterial cell. The DNA then becomes a permanent part of the cell.

TRANSFORMATION: Incorporation by bacteria into their own chromosomes of naked DNA from solution.

TRANSLOCATION (chromosomal): An error in chromosomal replication, whereby one chromosome or a fragment thereof becomes attached to another chromosome. Balanced translocations are those where there is no associated net gain or loss of genetic material in the cell. In unbalanced translocations, genes are either gained or lost during the rearrangement.

TRISOMY: The state in which a cell or an individual has three, rather than two, of any specific chromosome.

VIRUS: A short length of DNA surrounded by a protein coat, which exists as an intracellular parasite. Some viruses destroy their hosts, but others ("temperate viruses") simply integrate their DNA into the host chromosome and then join in the established cell life cycle, usually without obvious harmful effects. *(See* LYSOGENY).

References

Preface

Abelson, P. H. Anxiety about genetic engineering. *Science* 173 (1971):285.

Anderson, W. E. The future shock of genetic medicine. *Medical Opinion.* 4, 2 (1975):54-61.

Davis, B. D. Prospects for genetic intervention in man. *Science* 170 (1970):1279-1283.

Dobzhansky, T. Changing man. *Science* 155 (1967): 409-415.

Fletcher, J. Ethical aspects of genetic controls. *New England Journal of Medicine* 285 (1971):776-783.

————. *The Ethics of Genetic Control.* New York: Anchor Press/Doubleday, 1974.

Kass, L. Babies by means of *in vitro* fertilization: unethical experiments on the unborn? *New England Journal of Medicine* 285 (1971):1174-1179.

————. The new biology: what price relieving man's estate? Science 174 (1971):779-788.

Lederberg, J. Genetic engineering, or the amelioration of genetic defect. *Pharos* 34 (1971):9-12.

Motulsky, A. G. Brave new world? *Science* 185 (1974):653-663.

Ramsey, P. *Fabricated Man.* New Haven: Yale University Press, 1970.

Sinsheimer, R. Genetic engineering: the modification of man. *Impact of Science on Society* 20 (1970):279-291.

Introduction

Fechheimer, N. S. Causal basis of chromosome abnormalities. *Journal of Reproduction and Fertility Suppl.* 15 (1972):79-98.

Hook, E. B. Some comments on the significance of sex chromosome abnormalities in human males. In *Heredity and Society*, ed. I. H. Porter and R. G. Skalko. New York: Academic Press, 1973. Pp. 213-223.

Lilienfeld, A. M. *Epidemiology of Mongolism*. Baltimore: Johns Hopkins Press, 1969.

McKusick, V. A., and Claiborne, R., eds. *Medical Genetics*. New York: HP Publishing Co., 1973.

Montagu, A. *Human Heredity*. 2nd revised ed. Cleveland: World Publishing Company, 1963.

Sparkes, R. S., and Crandall, B. F. Genetic disorders affecting growth and development. In *Pathophysiology of Gestation*, Vol. II, Fetal-placental Disorders, ed. N. S. Assali. New York: Academic Press, 1972. Pp. 207-267.

Stern, C. *Principles of Human Genetics*. 3rd ed. San Francisco: W. H. Freeman and Company, 1973.

Thompson, J. S., and Thompson, M. W. *Genetics in Medicine*. 2nd ed. Philadelphia: W. B. Saunders Co., 1973.

Trimble, B. K., and Doughty, J. H. The amount of hereditary disease in human populations. *Annals of Human Genetics* 38(1974): 199-223.

Chapter 1

Borel, E. *Elements of the Theory of Probability*. Englewood Cliffs, N. J.: Prentice-Hall, 1965. (Chapters 1 and 2 reasonably comprehensible. After that, proceed at your own risk.)

Carter, C. O., Evans, K. A., Roberts, J. A. F., and Buck, A. R. Genetic clinic: a follow-up. *Lancet* 1 (1971):281-285.

Fraser, F. C. Genetic counseling. In *Medical Genetics*, ed. V. A. McKusick and R. Claiborne. New York: HP Publishing Co., 1973. Pp. 221-228.

Fraser, F. C. Genetic counseling. *American Journal of Human Genetics* 26 (1974):636-659.

King, A. C., and Read, C. B. *Pathways to Probability*. New York: Holt, Rinehart, and Winston, 1963.

Leonard, C. O., Chase, G. A., and Childs, B. Genetic counseling: a consumer's view. *New England Journal of Medicine* 287 (1972):433-439.

Wilson, A. A., and Smith, D. W. *The Child with Down's Syndrome*. Philadelphia. W. B. Saunders Co., 1973.

Chapter 2

Gordon, H. Genetics and civilization in historical perspective. In

Heredity and Society, ed. I. H. Porter and R. G. Skalko. New York: Academic Press, 1973. Pp. 3-43.

Gottesman, I. I., and Erlenmeyer-Kimling, L. A foundation for informed eugenics. *Social Biology* 18 Suppl. (1971):S1-S8.

Haller, M. H. *Eugenics: Hereditarian Attitudes in American Thought.* New Brunswick: Rutgers University Press, 1963.

Lederberg, J. Experimental genetics and human evolution. *American Naturalist* 100 (1966):519-531.

———. Biological innovation and genetic intervention. In *Challenging Biological Problems,* ed. J. A. Behnke. New York: Oxford University Press, 1972. Pp. 7-27.

Ludmerer, K. M. *Genetics and American Society.* Baltimore: Johns Hopkins Press, 1972.

Muller, H. J. Our load of mutations. *American Journal of Human Genetics* 2 (1950):111-176.

———. The guidance of human evolution. *Perspectives in Biology and Medicine* 3 (1959):1-43.

Penrose, L. S. Propagation of the unfit. *Lancet* 2 (1950):425-427.

———. Limitations of eugenics. *Proceedings of the Royal Institution of Great Britain* 39 (1962-63):506-519.

Smith, J. M. Eugenics and utopia. *Daedalus* 94 (1965):487-505.

Chapter 3

Beck, E., Blaichman, S., and Scriver, C. R. Advocacy and compliance in genetic screening. *New England Journal of Medicine* 291 (1974):1166-1170.

Bergsma, D., ed. *Ethical, Social and Legal Dimensions of Screening for Human Genetic Disease.* Miami: Symposia Specialists, 1974.

Buist, N. R. M. Neonatal screening tests for inborn errors of metabolism. *Contemporary Ob/Gyn* 4 (1974):59-62.

Childs, B. Prospects for genetic screening. *Journal of Pediatrics* 87(1975):1125-1132.

Culliton, B. J. Genetic screening: NAS recommends proceeding with caution. *Science* 189 (1975):119-120.

Guthrie, R. Mass screening for genetic disease. In *Medical Genetics,* ed. V. A. McKusick and R. Claiborne. New York: HP Publishing Co., 1973. Pp. 229-236.

Holtzman, N. A., Meek, A. G., and Mellits, E. D. Neonatal screening for phenylketonuria. 1. Effectiveness. *Journal of the American Medical Association* 229 (1974):667-670.

Kaback, M. M., and O'Brien, J. S. Tay-Sachs: Prototype for prevention of genetic disease. In *Medical Genetics,* ed. V. A. McKusick and R. Claiborne. New York: HP Publishing Co., 1973. Pp. 253-262.

Levy, H. L. Genetic screening. In *Advances in Human Genetics 4*, ed. H. Harris, and K. Hirschhorn. New York: Plenum Press, 1974. Pp. 1-104, 389-394.

Nitowsky, H. M. The significance of screening for inborn errors of metabolism. In *Heredity and Society*, ed. I. H. Porter and R. G. Skalko. New York: Academic Press, 1973. Pp. 225-261.

Chapter 4

Aposhian, H. V. The use of DNA for gene therapy — the need, experimental approach, and implications. *Perspectives in Biology and Medicine* 14 (1970):98-108.

Cohen, S. N. The manipulation of genes. *Scientific American* 233, No. 1 (1975):25-33.

Davis, B. D. Prospects for genetic intervention in man. *Science* 170 (1970):1279-1283.

Friedmann, T., and Roblin, R. Gene therapy for human genetic disease? *Science* 175 (1972):949-955.

Home therapy for haemophilia. *British Medical Journal* 1 (1975):417.

Howell, R. R. Genetic disease: the present status of treatment. In *Medical Genetics*, ed. V. A. McKusick and R. Claiborne. New York: HP Publishing Co., 1973. Pp. 271-280.

Lederberg, J. DNA splicing: Will fear rob us of its benefits? *Prism* (Nov. 1975):33-37.

Merril, C. R., Geier, M. R., and Trigg, M. E. Transduction in mammalian cells. In *Birth Defects. Proceedings of the Fourth International Conference*, ed. A. G. Motulsky and H. Lenz. New York: Elsevier Publishing Co., 1974. Pp. 81-91.

Rogers, S. Skills for genetic engineers. *New Scientist* 45 (1970):194-196.

Scriver, C. R. Enzyme therapy and induction in genetic disease: pox or pax. In *Birth Defects. Proceedings of the Fourth International Conference*. ed. A. G. Motulsky and H. Lenz. New York: Elsevier Publishing Co., 1974. Pp. 114-126.

Sinsheimer, R. Genetic engineering: the modification of man. *Impact of Science on Society* 20 (1970):279-291.

———. Troubled dawn for genetic engineering. *New Scientist* 68 (1975):148-151.

Swearing off gene transplantation. *Medical World News* (Aug. 9, 1974):18-19.

Wade, N. Genetics: conference sets strict limits to replace moratorium. *Science* 187 (1975):931-935.

Chapter 5

Burton, B. K., Gerbie, A. B., and Nadler, H. L. Present status of intrauterine diagnosis of genetic defects. *American Journal of Obstetrics and Gynecology* 118 (1974):718-746.

Friedman, J. M. Legal implications of amniocentesis. *University of Pennsylvania Law Review* 123 (1974):92-156.

Friedmann, T. Prenatal diagnosis of genetic disease. *Scientific American* 225, No. 5 (1971):34-42.

Golbus, M. S., Conte, F. A., Schneider, E. L., and Epstein, C. J. Intrauterine diagnosis of genetic defects: results, problems, and follow-up of one hundred cases in a prenatal genetic detection center. *American Journal of Obstetrics and Gynecology* 118 (1974):897-905.

Harris, H. Prenatal diagnosis and selective abortion. London: Nuffield Provincial Hospitals Trust, 1974.

Karp, L., Fialkow, P. J., Hoehn, H., and Scott, C. R. Prenatal diagnosis of genetic diseases. *University of Washington Medicine* 1 (1974):4-13.

Milunsky, A. The prenatal diagnosis of hereditary disorders. Springfield, Ill.: Charles C. Thomas, 1973.

Milunsky, A., Littlefield, J. W., Kanfer, J. N., Kolodny, E. H., Shih, V. E., and Atkins, L. Prenatal genetic diagnosis. *New England Journal of Medicine* 283 (1970):1370-1381, 1441-1447, 1498-1504.

Milunsky, A., and Reilly, P. The "new" genetics: emerging medicolegal issues in the prenatal diagnosis of hereditary disorders. *American Journal of Law and Medicine* 1 (1975):71-88.

Nadler, H. L. Prenatal diagnosis of inborn defects: a status report. *Hospital Practice* 10, No. 6 (1975):41-51.

Stone, M. L., Weingold, A. B., and Lee, B. O. Clinical applications of ultrasound in obstetrics and gynecology. *American Journal of Obstetrics and Gynecology* 113 (1972):1046-1052.

Thompson, H. E. The clinical use of pulsed echo ultrasound in obstetrics and gynecology. *Obstetrical and Gynecological Survey* 23 (1968):903-932.

Wheeless, C. R. Use of the laparoamnioscope for fetal observation and its effect on the human fetus. *American Journal of Obstetrics and Gynecology* 119 (1974):844-848.

Chapter 6

Behrman, S. J., and Kistner, R. W., eds. *Progress in Infertility*. Boston: Little, Brown and Co., 1968. Section 8, pp. 711-750.

Frankel, M. S. Role of semen cryobanking in American medicine. *British Medical Journal* 3 (1974):619-621.

Gerstel, G. A psychoanalytic view of artificial donor insemination. *American Journal of Psychotherapy* 17 (1963):64-77.

Guttmacher, A. F. The role of artificial insemination in the treatment of sterility. *Obstetrical and Gynecological Survey* 15 (1960):767-785.

Kleegman, S. J. Practical and ethical aspects of artificial insemination. In *Advances in Sex Research,* ed. H. G. Beigel. New York: Harper and Row, 1963. Pp. 112-118.

Muller, H. J. The guidance of human evolution. *Perspectives in Biology and Medicine* 3 (1959):1-43.

Sherman, J. K. Synopsis of the use of frozen human semen since 1964: state of the art of human semen banking. *Fertility and Sterility* 24 (1973):397-412.

Smith, K. D., and Steinberger, E. Survival of spermatozoa in a human sperm bank. *Journal of the American Medical Association* 223 (1973):774-777.

Steinberger, E., and Smith, K. D. Artificial insemination with fresh or frozen semen. *Journal of the American Medical Association* 223 (1973):778-783.

Strickler, R. C., Keller, D. W., and Warren, J. C. Artificial insemination with fresh donor semen. *New England Journal of Medicine* 293 (1975):848-853.

Weinstock, N. Artificial insemination — the problem and the solution. *Family Law Quarterly* 5 (1971):369-397.

Wolstenholme, G. E. W., and Fitzsimons, D. W., eds. *Law and Ethics of A.I.D. and Embryo Transfer.* Ciba Foundation Symposium 17 (new series). Amsterdam: Associated Scientific Publishers, 1973.

Chapter 7

Barlow, P., and Vosa, C. G. The Y chromosome in human spermatozoa. *Nature* 226 (1970):961-962.

Etzioni, A. Sex control, science, and society. *Science* 161 (1968):1107-1112.

Guerrero, R. Association of the type and time of insemination within the menstrual cycle with the human sex ratio at birth. *New England Journal of Medicine* 291 (1974):1056-1059.

Selecting the sex of one's children. *Lancet* 1 (1974):203-204.

Westoff, C. F., and Rindfuss, R. R. Sex preselection in the United States: some implications. *Science* 184 (1974):633-636.

Chapter 8

Edwards, R. G. Studies on human conception. *American Journal of Obstetrics and Gynecology* 117 (1973):587-601.

Edwards, R. G., and Fowler, R. E. Human embryos in the laboratory. *Scientific American* 223 (1970):44-54.

Edwards, R. G., and Sharpe, D. J. Social values and research in human embryology. *Nature* 231 (1971):87-91.

Edwards, R. G., Steptoe, P. C., and Purdy, J. M. Fertilization and cleavage *in vitro* of preovulatory human oocytes. *Nature* 227 (1970):1307-1309.

Grossman, E. The obsolescent mother. *Atlantic* 227 (May 1971):39-50.

Karp. L. E., and Donahue, R. P. Preimplantational ectogenesis. *Western Journal of Medicine,* 124(1976):282-298.

Kass, L. R. Babies by means of *in vitro* fertilization: unethical experiments on the unborn? *New England Journal of Medicine* 285 (1971):1174-1179.

Marx, J. L. Embryology: out of the womb — into the test tube. *Science* 182 (1973):811-814.

Mukherjee, A. B. Normal progeny from fertilization *in vitro* of mouse oocytes matured in culture and spermatozoa capacititated *in vitro*. *Nature* 237 (1972):397-398.

Standaert, T. A., Alden, E. R., Parks, C. R., Woodrum, D. E., Hessel, E. A., Murphy, J., Orr, R. J., and Hodson, W. A. Extracorporeal support of the fetal lamb simulating *in utero* gas exchange. *Gynecologic Investigation* 5 (1974):93-105.

Steptoe, P. C., and Edwards, R. G. Laparoscopic recovery of preovulatory human oocytes after priming of ovaries with gonadotrophins. *Lancet* 1 (1970): 683-689.

Westin, B., Nyberg, R., and Enhorning, G. A technique for perfusion of the previable human fetus. *Acta Paediatrica* 47 (1958):339-349.

Chapter 9

Austin, C. R. *The Mammalian Egg.* Oxford: Blackwell Scientific Publications, 1961.

Balfour-Lynn, S. Parthenogenesis in human beings. *Lancet* 1 (1956):1071-1072.

Graham, C. F. Parthenogenetic mouse blastocysts. *Nature* 226 (1970):165-167.

Graham, C. F., and Deussen, Z. A. *In vitro* activation of mouse eggs. *Journal of Embryology and Experimental Morphology* 31 (1974):497-512.

Kaufman, M. H., Huberman, E., and Sachs, L. Genetic control of haploid parthenogenetic development in mammalian embryos. *Nature* 254 (1975):694-695.

Parthenogenesis in mammals? *Lancet* 2 (1955):967-968.

Pincus, G., and Shapiro, H. Further studies on the parthenogenetic activation of rabbit eggs. *Proceedings of the National Academy of Sciences* 26 (1940):163-165.

Tarkowski, A. K., Witkowska, A., and Nowicka, J. Experimental parthenogenesis in the mouse. *Nature* 226 (1970):162-165.

Werthessen, N. T. with R. C. Johnson. Pincogenesis — parthenogenesis in rabbits by Gregory Pincus. *Perspectives in Biology and Medicine* 18 (1974):86-93.

Chapter 10

Briggs, R., and King. T. J. Transplantation of living nuclei from blastula cells into enucleated frogs' eggs. *Proceedings of the National Academy of Sciences* 38 (1952):455-463.

Gurdon, J. B. Transplanted nuclei and cell differentiation. *Scientific American* 219 (1968):24-35.

Halacy, D. S., Jr. *Genetic Revolution*. New York: Harper and Row, 1974. Chapter 11.

Kass, L. New beginnings in life. In *The New Genetics and the Future of Man*, ed. M. P. Hamilton. Grand Rapids: Eerdmans Publishing Co., 1972. Pp. 42-49.

Ladda, R. L., and Estensen, R. D. Introduction of a heterologous nucleus into enucleated cytoplasms of cultured mouse L-cells. *Proceedings of the National Academy of Sciences* 67 (1970):1528-1533.

Lederberg, J. Experimental genetics and human evolution. *American Naturalist* 100 (1966):519-531.

Sethi, K. K., and Brandis, H. Introduction of mouse L-cell nucleus into heterologous mammalian cells. *Nature* 250 (1974):225-226.

Thomas, L. On cloning a human being. *New England Journal of Medicine* 291 (1974):1296-1297.

Watson, J. D. Moving toward the clonal man — is this what we want? *Atlantic* (May 1971), pp. 50-53.

Chapter 11

Danielli, J. F. Artificial synthesis of new life forms. *Bulletin of the Atomic Scientists* 28 (Dec. 1972):20-24.

————. Context and future of cell synthesis. *New York State Journal of Medicine* 72 (1972):2814-2815.

Frankel, E. *DNA: Ladder of Life*. New York: McGraw-Hill, 1964.

Halacy, D. S., Jr. *Genetic Revolution*. New York: Harper and Row, 1974. Chapter 12.

Lessing, L. *DNA: At the Core of Life Itself*. New York: Macmillan, 1967.

Maugh, T. H. Molecular biology: a better artificial gene. *Science* 181 (1973):1235-1236.

Morowitz, H. J. Manufacturing a living organism. *Hospital Practice* 9 (1974):210-215.

Index

Achondroplasia, 38
Activation, of egg, 185
Adrenal hyperplasia, congenital, 124
Adriaenssens, K., 97
Allele, 5, 10, 24, 95
Alpha-fetoprotein, 118, 179
Amoebas, 196
Anemia, sickle cell, 42, 62, 72–75, 115, 122, 126
Anencephaly, 40–41, 79, 179
Aneuploid, 10, 18
Apoenzyme, 97–98
Artificial insemination, 131–47
    eutelegenesis, 144–47
    history, 132–35
    with husband's semen, 131
    legal problems, 138–40
    moral and ethical problems, 135–37
    psychological problems, 137–38
    semen banks, 140–44
    techniques, 132–35
Austin, C. R., 187

Barlow, Peter, 153
Barr body, 10
Behrman, S. J., 137
Bennett, D., 154
Bevis, D. C. A., 108, 174–75, 177
Bishop, D. W., 153
Blastocele, 163
Blastocyst, 163
Boyse, E. A., 154

Braden, A. W. H., 187
Brandis, H., 200
Brewer, Herbert, 144
Briggs, Robert, 197–98, 199
Burden, 40–41

Carter, C. O., 46, 53, 55
Celler Act, 52
Centromere, 5
Chang, M. C., 187
Character, 3
Chimera, 105, 171
Christ, 185
Chromosome, 5, 7, 15, 21–24, 39, 80, 114–15
    autosomal, 7
    definition, 5
    homologous, 5
    replication, 15
    sex, 7
    translocation, 21–24, 39, 80 114–15
Cleft lip and palate, 40, 93
Clone, 195
Clonee, 196
Cloning, 59, 195–206
    clinical applications, 203–6
    controversies, 203–6
    definitions, 195–96
    by embryo fission, 200–201
    by nuclear transfer, 196–200
    research applications, 201–3
Coenzyme, 97–98

Cold shock, 141
Colon, hereditary polyposis, 92–93
Compensation, reproductive, 56
Consanguinity, 27, 31, 37, 58
Coolidge, Calvin, 52
Coronary heart disease, 75–76
Corpus luteum, 163, 165, 172, 173, 178
Creatine phosphokinase, 77
Crew, James, F., 53, 54, 60
Cullen, Thomas, 41
Cystic fibrosis, 62, 115–16
Cytochalasin B., 199–200
Cytomegalovirus, 124–25

Danielli, James, 210
Davenport, Charles, B., 51
De Kretzer, D., 174
Di Berardino, Marie, 199, 202
Diabetes, 61, 77–79, 96–97
Diasio, R. B., 155
Dickinson, Robert L., 132
Dickson, Robert C., 208
Differentiation, cellular, 196–97
Disease
  congenital, 15
  familial, 15
  genetic, 5, 15, 18–32
    autosomal dominant, 24–25
    autosomal recessive, 25–27
    chromosomal, 18–24
    multifactorial, 31–32
    single gene, 24–31
    X-linked, 27–31
Donahue, Roger, 170
Down's Syndrome, 15, 18–19, 22–24,
    35, 39, 43, 107, 113–14, 179, 183
Dwarfism, diastrophic, 38, 61

Ectogenesis, 161–84
  definition, 161
  fertilization to implantation,
    170–75
  potential applications, 177–79
  moral and ethical issues, 180–84
  postimplantational, 175–77
  potential applications, 179–80
  prefertilization, 166–70

potential applications, 177–79
  research applications, 179
Edwards, Robert G., 170, 172–74, 177
Electrophoresis, hemoglobin, 73
Embryogenesis, natural, 162–65
Enhorning, G., 175
Enzmann, E. V. 170
Enzyme, 3
Epilepsy, 82, 90–91
Episome, 101
Ericsson, R. J., 153
Erlenmeyer-Kimling, 54
Estensen, Richard, 199–200
Eugenics, 49–60
  controversial aspects, 54–59
  historical aspects, 49–54
  negative, 50
  positive, 50
Euphenics, 62–63
Euploid, 10
Eutelegenesis, 144–46
Euthenics, 60–62
Evans, H. J. 153
Evans, M. Stanton, 183

Fabry's Disease, 97, 98
Fertilization
  natural, 162–65
  in vitro, 58–59, 170–75
Fetoscopy, 125–27
Fitness, genetic, 15
Fletcher, Joseph, 45, 180
Follicle cells, 163
Folling, Ashborn, 66–67, 68
Fraser, George, 60

Galactosemia, 70, 89–90, 102, 124
Galton, Francis, 49, 50, 51
Garrod, Sir Archibald, 35
Gaucher's Disease, 97
Gene, 3, 207–8
  carrier, asymptomatic, 27, 75–76,
    122, 159
  deletion, 21, 24
  de-repression, 105
  dominant, 10
  duplication, 21, 24

linkage, 58
recessive, 10
Genetic counseling, 35-48
    controversial aspects, 41-45
    definition, 36
Genetic screening, 65-83
    autosomal dominant disorders,
        75-76
    autosomal recessive disorders,
        66-75
    for carrier detection, 70-72
    chromosomal disorders, 79-80
    controversies, 80-83
    for early diagnosis, 66-70
    for early diagnosis and carrier
        detection, 72-75
    generalized, 72, 81-82
    multifactorial disorders, 77-79
    risks, 82-83
    selective, 72, 81-82
    X-linked recessive disorders, 76-77
Genetic therapy, 85-106
    avoidance of dangerous substances,
        94-95
    correction of defective genes,
        99-105
    dietary, 89-90
    drug therapy, 90-92
    enzyme therapy, 96-98
    surgery, 92-94, 98
German, James, 160
Gerstel, Gerda, 137-38, 145
Glass, R. H., 155
Glucose-6-phosphate dehydrogenase,
    76, 94-95
Glycogen storage diseases, 93-95
Gonadotrophin, 162
Gottesman, I. I., 54
Graham, C. F., 187, 200
Grant, Madison, 51-52
Guerrero, Rodrigo, 156, 160
Gulyas, B. J., 192
Gurdon, J. B., 198, 199, 201
Guthrie, Robert, 68, 81
Guthrie test, 68-69
Guttmacher, A., 132, 133, 135, 136.
    137

Haploid, 11
Hatzold, Otfried, 156-57, 159
Heart, congenital abnormalities, 93
HeLa cells, 20
Hemizygous, 110
Hemophilia, 27, 28-34, 76, 96-97, 105,
    115, 119, 126, 151, 159, 208
Hertwig, Oskar, 185
Heterozygous, 10
Holoenzyme, 97-98
Homozygous, 10
Hormone, 96
Hsia, Y. E., 48
Hsu, Yu-chih, 176-77
Human chorionic gonadotrophin,
    172-73, 186-87
Huntington's chorea, 15, 44
Hyperargininemia, 102-3
Hypertension, 77-79

Index case. *See* Propositus
Inner cell mass, 163
In vitro, 165-66
In vivo, 165-66

Jenner, Edward, 102

Kaback, Michael, 71-72, 81
Karyotype, 5, 39
Kass, Leon, 45, 161, 180, 182, 183, 184
Kennedy, Joseph, 170
Khorana, Har Gobind, 208
King, Thomas, 197-98, 199
Kleegman, S., 132, 135, 137, 156, 160
Klinefelter's Syndrome, 21, 79, 82,
    90-91
Kolstad, Per, 45
Kornberg, Arthur, 208
Ladda, Roger, 199-200
Langevin, C. N., 153
Lappé, Marc, 181
Laughlin, Harry, 51-52
Lederberg, Joshua, 62-63
Leo, 135
Leonard, C. O., 46
Leonard, Sheldon, 35
Life, synthesis of from nonliving

Life *(continued)*
    matter, 207–14
  DNA synthesis, 207–9
  human synthesis, 212
  metazoa, nonhuman synthesis, 211–12
  prokaryotic cell synthesis, 209–10
  protozoan synthesis, 210–11
  superhuman synthesis, 213–14
Liley, A. W., 108
Lillie, F. R., 185
Living organism, 207
Loeb, Jacques, 185, 186
London, Jack, 51, 57
Lowry, R. B., 38
Luteinizing hormone, 162, 166, 170, 172, 186
Lynch, H. T., 46
Lysogeny, 101

Marfan's Syndrome, 38
Maturation, oocyte, 166–70
McKusick, Victor, 38
Media, tissue culture, 166, 169
Meiosis, 11, 162, 163, 166, 186, 188, 195
Mencken, Henry L., 51
Mendel, Gregor, 49, 50, 51
Metabolism, inborn errors, 27, 35, 37–38
Mintz, Beatrice, 200–201
Mitosis, 15
Mongolism. *See* Down's Syndrome
Monosomy, 20–21
Morowitz, Howard J., 209–10
Morula, 163
Moses, 138
Motulsky, Arno, 60
Mukherjee, Anil, 171–72, 177
Muller, H. J., 53, 61, 62, 144, 145, 204
Murray, R. F., 46
Muscular dystrophy, 27, 76, 115, 119, 126, 151
Mutation, 3
Myopia, 61

Nadler, Henry, 107, 108
Namath, Joe, 58–59

National Genetics Foundation, 46
Nietzsche, Friedrich, 50
Nishino, M., 153
Nixon, Richard, 145
Nondisjunction, 24
Noyes, John Humphrey, 50
Nyberg, R., 175

Oneida Community, 50
Orotic aciduria, 90
Ovulation, natural, 162–65

Parthenogenesis, 185–93
  definition, 185
  experimental, 185–88, 190–93
  in humans, 190–91
  possible applications, 191–93
  spontaneous, 188–90
Pearson, Karl, 50, 60
Pedigree, 37, 39
Penrose, Lionel, 53, 55, 60
Pergonal, 172
Peyser, Herbert, 137, 145
Pharmacogenetics, 94–95
Phenylketonuria, 66–70, 89, 115
Pincus, Gregory, 170, 185–87, 192
Pituitary gland, 162
Placenta, artificial, 175–76
Pleiotropy, 58
Poland, B. J., 38
Porphyria, acute intermittent, 94–95
Prenatal diagnosis, 107–27
  amniocentesis, 108–18
  anencephaly, 116–18, 119
  chromosomal diseases, 113–15
  controversies, 120–23
  future developments, 124–27
  historical aspects, 108
  inborn errors of metabolism, 115–16
  spina bifida, 116–18
  thrombocytopenia-absent radius syndrome, 120
  translocations, 114–15
  ultrasound, 118–20
  X-linked diseases, 115
  X-ray, 120
Privacy, invasion of, 43–45

Progesterone, 163, 165, 172, 173, 178
Propositus, 38, 40
Pseudo cholinesterase deficiency, 95

Ramsey, Paul, 45, 161
Rent-a-womb, 183
Reproductive engineering, 129–214
    definition, 131
Retinoblastoma, 92
Rh disease, 62, 105–6
Rindfuss, Ronald, 158
Risk, empiric, 32, 39–40
Risk, specific, 39
Roberts, A. M., 152
Robinson, Hugh, 126
Robinson, J. A., 153
Rothschild, Lord, 153
Rubella, 124–25

Sachs, Bernard, 70
Scrimgeour, J. B., 125
Sendai virus, 200
Sethi, K. K., 200
Sex chromatin, 10, 153, 178
Sex determination, 149–60, 178, 193
    empirical methods, 155–57
    general and historical, 149–52
    possible uses, 159–60
    postconceptual, 150–51
    potential problems, 157–59
    preconceptual, 151–52, 157
    sperm separation, 152–55
Sex ratio, 150, 157–59
Shah of Iran, 149–150
Shapiro, Herbert, 187
Shelley, Mary, 161
Shettles, Landrum, 152–57, 159, 160, 174
Shirodkar, V. N., 182
Smallpox, 102
Smith, H. Allen, 184
Social Darwinism, 51
Soranus, 108
Spencer, Herbert, 51
Spiegelman, Sol, 208
Spina bifida, 40–41, 79, 116–18, 125, 179
Spurway, Helen, 190–91

Stalin, Joseph, 145
Steptoe, Patrick, 172-74, 177
Stokes, W. E. D., 58
Sumner, A. T., 153
Surrogate uterus, 183
Syndrome. *See* Down's Syndrome; Klinefelter's Syndrome; Marfan's Syndrome; Turner Syndrome; Waardenburg's Syndrome

Tarkowski, A., 187–88
Tay, Warren, 70
Tay-Sachs Disease, 37, 70–72
Thalassemia, 37
Thomas, Lewis, 204
Toxoplasma, 124–25
Trait, 3, 57
Transduction, 99–104
Transformation, 99
Trisomy, 18–21
Trophoblast, 163
Turner Syndrome, 20–21, 79–80

Uterus, artificial, 176–77

Valenti, Carlo, 107, 108, 125
van Duijn, C., Jr., 153
Van Gogh, Vincent, 205
Virgin birth. *See* Parthenogenesis
Virus, 99–104, 207–9
    herpes, 101
    temperate, 101
Von Hoffman, Nicholas, 58–59, 161, 193
Vosa, C. G., 153

Waardenburg's Syndrome, 47
Westin, B., 175
Westoff, Charles, 158
Wheeless, C. R., 125
Whitten, W. K., 187, 190
Wilson's Disease, 44, 45, 90–92

XY female, 43

Yale Law Journal, 138

LAURENCE E. KARP is co-director of the Prenatal Diagnosis Center of the Department of Obstetrics and Gynecology, Division of Medical Sciences of U.C.L.A., located at Harbor General Hospital in Torrance, California. There, his interest in reproductive genetics has deeply involved him in research on the origins of chromosomal diseases as well as in the teaching of resident physicians, interns and medical students in his capacity as associate professor.

Dr. Karp obtained his M.D. from New York University School of Medicine. After graduation he served as coordinator of the Prenatal Diagnosis Clinic, Center for Inherited Diseases, at the University of Washington School of Medicine.

Medical and scientific articles authored by Laurence Karp have appeared in the *Smithsonian, Medical Dimensions,* the *New York Times* "Op-Ed" pages, and in numerous professional journals. He is a fellow of the American College of Obstetricians and Gynecologists and a member of the American Society of Human Genetics and the American Association for the Advancement of Science.

Can we produce 10,000 Mao Tse-Tungs by cloning? Will humans be replaced by nonhuman forms of life? Are geneticists defying their Maker and usurping divine prerogatives? Are they leading us back to the Nazi or ahead to *Brave New World*? What about test-tube babies? Robots? Human-animal chimeras? Specially bred legless astronauts?

These questions indicate some of the threats of genetic engineering and they are sensational. Science fiction lovers —acknowledged and unacknowledged—thrill to these threats in print and on the screens: Stop the mad geneticist or he will lead us down the road to atrocity, dehumanization, and slavery.

We hear less about the promises of genetic engineering:

- breakthroughs in preventing birth defects
- knowledge to alleviate other human genetic disorders, heart disease, and even cancer
- means to feed the millions of hungry people in our world.

Genetic engineering is defined here as "the art or science of making practical application of the knowledge of the pure science of genetics." Genetic engineers are concerned with any attempt to modify the structure, transmission, expression, or effect of the ultimate directors of heredity—the genes. Specific areas are:

- artificial insemination